ECOLOGICAL CIVILIZATION OF CONTEMPORARY CHINA

By Cao Baoyin & Yin Wujin

 China Intercontinental Press

图书在版编目（CIP）数据

当代中国生态文明：英文/曹保印，尹武进著；译谷译. -- 北京：五洲传播出版社，2014.6（当代中国系列/武力主编）

ISBN 978-7-5085-2785-7

Ⅰ.①当… Ⅱ.①曹…②尹…③译… Ⅲ.①生态环境建设－概况－中国－英文 Ⅳ.① X321.2

中国版本图书馆 CIP 数据核字 (2014) 第 124452 号

当代中国系列丛书

主　　编：武　力
出 版 人：荆孝敏
统　　筹：付　平

当代中国生态文明

著　　者：曹保印　尹武进
译　　者：译　谷
责任编辑：王　峰
图片提供：中新社　CFP　东方 IC　曹保印　尹武进
装帧设计：丰饶文化传播有限责任公司
出版发行：五洲传播出版社
地　　址：北京市海淀区北三环中路 31 号生产力大楼 B 座 7 层
邮　　编：100088
电　　话：010-82005927，82007837
网　　址：www.cicc.org.cn
承 印 者：中煤涿州制图印刷厂北京分厂
版　　次：2014 年 6 月第 1 版第 1 次印刷
开　　本：787×1092 毫米 1/16
印　　张：11.75
字　　数：150 千字
定　　价：108.00 元

Contents

Foreword ——————————————————————— 5

Ecological Civilization and Chinese Dream ———————— 24

    Respecting Nature: Story of a Black-tailed Gull
Called "Tianci" (Gift of Nature) ————————————————— 28

    Conforming to Nature: Explore the "Chinese-style Environmental Protection" ——————————————————————————— 32

    Protecting Nature: Chinese Environmental Protection Dream ———— 38

Blue Sky ——————————————————————————— 44

    PM2.5: A Reformation Caused by a Cough ———————————— 48

    RMB 5 Trillion for 5 Years: The Capital City Surrounded
by "Haze" ——————————————————————————— 54

    Urban Ventilated Corridor: Hangzhou's Actions
of "Introducing Wind and Dispersing Haze" ——————————— 61

    Thorough Treatment of Pollution in Lanzhou ———————————— 66

    Industrial Relocation: Capital or Shougang? ———————————— 71

    Joint Prevention and Control: "Pilot Pollution Control" Program
of Beijing-Tianjin-Hebei ———————————————————— 76

Green Land — 84

Great Changes in Coal City: "Coal Sea" Changing
Into Sea of Trees — 88

Hero in the Desert: "Shihuichui" Fighting Against Langwosha
for Three Times — 92

Grassland Restoration Experiment:
The "Wise Method" of Jiang Gaoming — 98

Epitaph in a Wet Land: The Loss and Save of "Eden" — 103

Income of RMB 11 Billion from Furs: Behind the Fur Capital — 108

Clear Water — 114

Cancer Village: Death Village under the Shadow of Pollution — 119

RMB 8 Billion of Salvation: Truth of Dead Fish in Baiyangdian Lake
"Surfaces above the Water" — 124

Public Benefit Database: the Wisdom of Water Pollution Map — 129

Ecosystem Assessment: "Eight Doubts" for Lake-lining Project
of the Old Summer Palace — 133

Ten Years of Protection of Nujiang River of Wang Yongchen — 141

Pollution Control for Huaihe River: The Shocking "Thirty-Six
Strategies" for Illegal Pollution Discharge — 148

Beautiful China — 156

Green Development: Green Strategy of 3D Printing — 160

Cyclic Development: The Lights-out Earth Hour — 165

Low-carbon Development: Green GDP of Hezhang — 171

Postscript — 178

Foreword

"We will resolutely declare war against pollution as we declared war against poverty."

Obviously, the confidence, determination, ambition and the pragmatic attitude of the declaration have caused a strong emotional and value resonance with all Chinese. The person saying these words shouted the aspirations that every Chinese has been looking forward to for a long time, which also reflects the reality of urgency to solve pollution problem.

Who said those words at what time in what occasion so categorically vowing to treat pollution with iron hands? It's Li Keqiang, prime minister of the State Council of People's Republic of China. On the morning of March 5, 2014, in the Great Hall of the People in Beijing, at the opening ceremony of the Second Session of 12th National People's Congress, Li Keqiang showed his determination to treat pollution when making report on the work of the government on behalf of the current central government.

As he put it, "Fostering a sound ecological environment is vital for people's lives and the future of our nation. Smog is affecting larger parts of China and environmental pollution has become a major problem, which is nature's red-light warning against the model of inefficient and blind development. We must strengthen protection of the ecological environment and resolve to take hard measures to complete this hard task."

Premier Li Keqiang has given the most authoritative answer. With two words of "hard", he expressed his uncompromisingly great courage and determination. Therefore, we can understand how arduous China's ecological civilization construction is and how serious the contradiction of the environmental pollution is so that it won't be solved without the hard measures.

Just viewed from the aspect of energy consumption, let's look at a simple set of data: Total primary energy consumption in China in 2012 was equivalent to 3.62 billion tons of standard coal, accounting for 21.3% of the global total. The consumption of unit GDP was as twice as that of the world, and 4 times as that of the developed countries. Even if the "Twelfth Five-year Plan" controls the GDP growth around 7.5%, the estimated total primary energy consumption will be 4 billion tons of standard coal in 2015 and 4.5 billion tons in 2020.

If we keep going the way of "extensive development", China's resources and the environment will not be sustainable. This also indicates that despite of continuous decline of energy consumption and somewhat reduction of total amount of pollution discharge due to the strict assessment indicators implemented from national to local governments and from industry to enterprise, the rapid development of economic activities is exerting more and more pressure on resources and environment; the industrial growth remains over-reliance on investment in material resources; and the deterioration of the ecological environment has not been effectively curbed.

It is the Chairman of China Federation of Industrial Economics Li Yizhong, rather than others, who drew above conclusion and clearly expressed his concerns. On July 30, 2013, at the "Caring for Climate China Summit" hosted

Foreword

Li Yizhong, Chairman of the China Federation of Industrial Economics, deeply concerning about China's ecological environment problems.

by Global Compact Network China, outspoken Li Yizhong frankly and sharply expressed his above opinion.

The opinion that "ecological environment deterioration hasn't been effectively curbed" had already been expressed by a more authoritative institute 7 years before Li Yizhong made his speech.

The day of June 5, 2006 was the 35th World Environment Day. On that day, the former State Environmental Protection Administration published the document *Nature and Ecology Conservation in China* for the first time, which made a comprehensive introduction to ecology conservation in China. In this authoritative document, the related responsible person of the administration clearly pointed out that the government of China has attached great importance to ecological environmental protection and construction, taken a series of strategic measures and intensified the efforts on ecological environment protection and construction, and the environment in some key regions has seen improvement, but the per capita resources in China is in short and varies from place to place, the ecological environment is fragile, and "ecological environment deterioration hasn't been effectively curbed".

But how exactly serious is the environment deterioration in China?

Let's have a look at another set of data: In 2000, the national environmental quality evaluation result showed that nearly one third of land of China was good in ecological environmental quality and another one third was in poor or worse condition. The 6th national forest assessment results showed that the forestry area and the growing stock got increased and the forest quality got improved. But the problems of shortage of total resource amount, unequal distribution, low quality, and over-exploitation still existed; natural prairie of China shared 41% of the national territorial area, but 90% of it suffered from degeneration and desertification, which has been the main reason of windy and dusty weather; wetland area of China ranked the first in Asia, and fourth in the world, and 40% of it were under good protection, but serious reduction, degeneration and disappearing of wetland were still regrettable. Arable land accounts for 12.7% of China's land area, and it is an important component of terrestrial ecosystem. But during agricultural production process, slather of chemical fertilizers, pesticides, agricultural film have some negative influence on agricultural productivity and the surrounding natural ecosystems. China's sea area is of about 1/3 of land area. But due to rapid population growth in coastal areas and rapid economic development, coastal beaches, wetlands, ecological destruction has been intensified, and the overall pollution of waters has not been improved. China is one of the countries with the largest desert, which is mainly located in the northwest arid area. Excessive use of plant and over-exploitation of upstream water resources of inland rivers have led to ecological deterioration and desertification. Chinese urban green area continues to expand, but problems of shortage of water resources in the city and small urban green area and poor function still exist. The "dirty, chaotic, and poor" phenomena still exist. Issues of

Foreword

In the south of Qinghai Province, a horse is walking on a dry riverbed at the drought-hit source of Yangtze River.

agricultural non-point source pollution and livestock pollution remain serious.

Land of fragile ecological environment shares more than 60% of the total area. Ecological environment pressure is prominent, for Chinese per capita resources is less than half of the world average, but the unit GDP energy consumption and material consumption are much higher than the world average. Ecological protection policies, regulations and standards still need to be improved. Although China has promulgated a series of environmental protection and natural resource management laws and regulations, these laws have different emphases, resulting in the lack of a comprehensive ecological protection law. Insufficient investment in ecological protection, in capital input, and single channel show that effectiveness of eco-governance project still need to be improved. The technology and information support of ecological protection are weak, ecological protection research power is limited, leading to the difficulty to support the management.

Ecological Civilization of Contemporary China

An abandoned county at the source of Yellow River due to lack of water.

After 30 years of rapid development, China has become the world's second largest economy the accumulated environmental problems have emerged one after another as well, and the overall environmental deterioration has increased the pressure. As GDP grows, the huge "ecological deficit" appears. In the face of frequent environmental events, the single assessment indicator of economic growth is not functional. Data show that China has the largest consumption of energy, steel, and cement. In 2013 the total energy consumption got 3.75 billion tons of standard coal; among the 704 water quality monitoring sections in ten drainage basins, and inferior V-Class sections accounted for 8.9%. Serious pollution is a threat to the drinking water safety. And "cadmium rice" incident arouses the concern about the soil contamination problem...

A direct evidence: Around February 24, 2014, the satellite remote sensing monitoring of Ministry of Environmental Protection showed that more than one million square kilometers of land in China suffered from the dust-haze pollution.

For example, on February 22, 33 out of the 161 cities covered by the air quality new standard monitoring activity encountered intense pollution, and 10 cities worse. The pollution was primarily in Beijing-Tianjin-Hebei and surrounding regions, and central, western and northeastern China. Among them, Beijing-Tianjin-Hebei and surrounding regions suffered the most, with primary pollutants including PM2.5 and PM10.

On February 21, the air pollution emergency headquarter of Beijing issued the air pollution orange alert for the first time. According to the unified arrangement of emergency operation for heavily polluted weather, the industrial system carried out the *Heavy Air Pollution Industrial Emergency Subplan*. 36 enterprises stopped production and 75 reduced production. At zero o'clock on February 23, Shijiazhuang City of Hebei launched the emergency response to heavy pollution Grade II (orange) alert, restricting 20% of traffic within the third ring road, excluding vehicles of army, police, emergency treatment, rescuing,

A civilian, wearing a gauze mask, walks the dog in a day of heavy pollution for which orange warning was given by relevant Beijing authorities.

people's livelihood ensuring vehicles, buses, taxes, etc. During those days, the author of this book was living and working in Beijing, feeling deeply the negative impact of the air pollution on people's physical and psychological health.

The former Minister of Health Chen Zhu published the article entitled *China Tackles the Health Effects of Air Pollution* on the Lancet, the authoritative international medical magazine in December 2013, believing that the outdoor air pollution causes 350,000-500,000 people's premature death in China every year.

Ministry of Environmental Protection issued the first true air "annual report" on March 25, 2014. The results were shocking: Air quality in 71 out of the 74 large and medium cities in China in 2013 failed to reach the standard, a share of 95.9% of the total. The qualified three are Haikou, Zhoushan and Lhasa. Around 300 million Chinese are living in those 74 cities.

The pollution brings hugely negative impact on China's economic, social development and people's livelihood. It forces us to declare war against the air pollution, or the consequences would be unthinkable. Against the very backdrop, Premier Li Keqiang sets his heart on declaring war against the pollution.

Since the war is declared, then next, how will we fight the pollution in the "hard" war? How will we win and curb the trend of ecological environment deterioration to construct an ecological China?

Premier Li Keqiang gives the "prescription": Hit the air pollution a hard blow and intensify the treatment and prevention. Concentrate on the haze-prone large cities and regions, take the treatment of the fine particulate matter (PM2.5) and the inhalable particle (PM10) as a breakthrough, focus on the industrial structure, energy efficiency, exhaust emissions and dust, and other key areas, improve the new mechanism jointly participated by government, corporate, public, carry out the regional joint prevention and control, and thoroughly implement the air pollution control action plan. In 2014, we will obsolete 50,000 small coal-fired boilers, conduct 15 million kilowatts of desulfurization, 130

Foreword

Green hydropower generated in the Wasi River, a branch of Dadu River in Ganzi Tibetan autonomous prefecture of Sichuan province.

million kilowatts of denitrification, and 180 million kilowatts of dedusting in coal-fired power plants; eliminate 6 million "yellow label cars" and old cars; provide the diesel fuel for motor vehicle of national standard four across China. Implement the clean water action plan to strengthen the protection of drinking water source, and promote pollution treatment in key watersheds. Start the project of soil restoration. Treat the agricultural non-point source pollution and construct our beautiful hometown.

The "prescription" is great, but how will we ensure the implementation?

Efforts of only the premier are obviously not enough. China needs a "national environment protection war", and every Chinese is required to "take up arms" and fight the pollution. In other words, we have already got the "green premier", what we need now are the "green citizens". Based on the consensus of "declaring war against the pollution", changing development pattern, preserving the clear water and blue sky, and constructing beautiful China can be the urgent tasks of Chinese party and government, and the common responsibility of all Chinese.

As a matter of fact, the report of 18th National Congress of the Communist Party of China has included the ecological civilization construction in the general plan of socialist modernization, put forward the policy of insisting on the priority of energy saving, protection and recovery of nature, and called for powers to make a beautiful China. To ensure the priority of protection, the philosophy that protecting the ecological environment is protecting the productivity and improving the ecological environment is developing productivity should be set up. And we should consciously contribute to the green development, cyclic development and low-carbon development, and never develop the economy at the cost of environment. It is worth mentioning that, the priority of environment protection embodies the deepened knowledge of the natural law and law of economic society development and the more scientific understanding of the relationship between development and environment protection of Chinese party and government.

Later on, the Third Plenary Session of 18th Central Committee of CPC proposed that, while concentrating on construction of a beautiful China, we should deepen the transform of ecological civilization system, accelerate the establishment of ecological civilization system, and improve the mechanism of land space exploitation, economical resources utilization, and ecological environment protection. The Third Plenary Session of 18th Central Committee of CPC has determined the cancelation of assessment of regional total production value in the key counties included in the national plan for poverty alleviation through development, where development is restricted and ecology is fragile, as well as the audit on the leaders in regard to natural resources assets after they have left the post, and the establishment of lifelong accountability system of ecological environment damage. The Organization Department of CPC Central Committee has issued the *Circular on Improving the Government Performance*

Foreword

Examination of the Local Party and Government Leadship and Leading Cadres, having delivered the clear expectation and requirements to improve the government performance assessment and evaluation standard, and cancel the rankings of gross regional production and growth rate. While explicitly canceling the assessment, the circular also demands to implement the performance evaluation on the work of making agriculture and ecological protection the top priority in the main agricultural products production area and key ecological functional areas where the development is limited, and to highlight the assessment of the achievement of poverty alleviation and development in the key counties included in the national plan for poverty alleviation through development.

Half a year before the Third Plenary Session of 18th Central Committee of CPC, on May 24, 2013, the General Secretary of CPC Central Committee Xi Jinping has emphasized in the sixth collective study of Politburo the correct handling of the relationship between economic development and ecological environment protection, and never develop the economy at the cost of environment. On environmental protection, lifelong accountability system has been put forward: Whoever crosses the "ecological red line" should be punished; whoever causes serious consequences must be held accountable, "and should be held accountable for life."

As emphasized by Xi Jinping, land is the spatial carrier of ecological civilization construction. Following the principle of achieving balance among population, resources and environment, and unifying the economic, social and ecological benefits, we should make the overall plan for land spatial development, scientific layout of production space, living space, and ecological space, to leave the nature with larger space for repair. Unwaveringly accelerate the implementation of the main functional area strategy, strictly accord with the main function orientation of optimizing development, emphasizing development, restricting development, and prohibiting development, designate and strictly

Ecological Civilization of Contemporary China

The exploration work at collapsed riverside caused by digging.

maintain the ecological red line, build the scientific and rational urbanization pattern, agricultural development pattern, and ecological security pattern, protect the national and regional ecological security, and improve the ecosystem services. We should firmly set up the concept of ecological red line. In the aspect of ecological environmental protection, anyone stepping over the line would be published.

Xi Jinping's words are the "most forceful statement" on ecology among those of Chinese leaders, and the speech "to declare war against air pollution" of Li Keqiang is the "most vigorous enforcement". The lifelong accountability system for ecological environment protection plays a very important role, because many officials used to strive for the short-term economic benefit or their own interests at the expense of environment. The negative impact on the ecology doesn't show immediately, but will be assumed by the successor. And environmental protection cannot get immediate results, so incumbents have little incentive to protect the environment. Therefore, it is also difficult for the current

Foreword

decision-makers to improve and strictly enforce the environmental protection laws and regulations. But in the face of lifelong accountability system, officials must go all out.

It can be seen that, driven by the great pressure of "lifelong accountability system", the government of Beijing, Yunnan, Hubei, Guangxi, Fujian, Zhejiang, Jiangsu, Guangdong, Chongqing, Shandong, Hainan and other provinces (autonomous regions and municipalities) have issued special documents in succession, placing the construction of ecological civilization in an important position where it can promote regional scientific development, taking the construction of ecological civilization as the important starting point to achieve overall regional economic development, social progress and environmental protection in a balanced way, to develop a systematic plan for ecological civilization construction.

In Chongqing, considering the difference of the city's economic

Yangtze River's First Beach, located to the north-west of Yunnan Province. likely to be submerged if the Tiger Leaping Gorge Dam is to be built here.

development level, natural resources, population carrying capacity, the city is divided into "five functional areas", namely, the urban functional core area, the urban functional expansion area, the urban development new district, the northeastern Chongqing ecological conservation and development area and the southeastern Chongqing ecological protection and development area. And it has the corresponding government performance evaluation and fiscal and financial protection mechanisms in place to promote regional differentiated and synergetic development. To ensure the implementation of strategies in the functional areas, performance appraisal systems of difference focuses are under establishment in Chongqing.

Hainan has determined to define the production, life and ecology development and regulation borders, to improve the natural resources property order and utilization management system, and to implement the most stringent measures to protect natural resources. Resources and environment carrying capacity monitoring and early warning mechanisms have been established,

Network drainage of the Yangtze River source.

and the regional restrictive measures on soil and water resources, environment capacity and in marine resources overload zones are taken.

The government of Beijing has proposed to deepen the ecological civilization system reform, innovate the urban management system, innovate the regional ecological environment cooperation mechanism, and solve the problem of environmental construction in the Capital. The key air pollution prevention and treatment regions including Beijing and Hebei have issued the implementation program for the air pollution and treatment action plan in the first place.

Guizhou has determined to take 7 reform measures, including exploring to prepare the natural resources balance sheet, launching the experimental audit on leading cadres' responsibility in regard to natural resource assets, promoting environmental tax reform, expanding eco-compensation reform pilot area, and exploring the establishment of lifelong accountability system for ecological damage. Guiyang City even has issued the trial implementation measures and included the indicators of ecological civilization into the assessment of government departments and officials across the city. Whether the environmental protection work reached the standard becomes an important basis for assessment, appointment and dismissing, rewarding and punishment of the cadres. The "environmental protection related governance" forced the government performance examination transfer to "green", which is of great significance.

At a higher level, in June 2013, executive meeting of the State Council deployed ten air pollution control measures; in September, the State Council promulgated the Air Pollution Prevention Action Plan, putting forward the specific objectives: By 2017, concentration of inhalable particulate matters in the prefecture-level and above cities would decrease by more than 10% compared with 2012, and the number of days with good air quality gradually increase, etc. In the meantime, water environment comprehensive improvement in the important lakes, pollution control in key river valleys, and construction of beautiful countryside are carried out in an orderly way in the country.

In addition to those specific practices, at the higher level of the system designing and system guaranteeing, the construction of beautiful China is gradually becoming a reality. Because only by adhering to the basic national policy of conserving resources and protecting environment, and supported by the most strict system and the most stringent rules, can the ideas be carried out to ensure the lifelong accountability system and enter a new era of Chinese ecological civilization construction. And only in that way can the beautiful China be built, leaving a blue sky for future generations.

For example, establish and improve the ecological protection laws, regulations and standard systems, include environmental protection and construction into the national legalized management system, and intensify the investigation and punishment of major ecological damages in key areas and watersheds. The draft amendment to the environmental protection law submitted to and deliberated by the NPC Standing Committee clearly states: "Protecting the environment is a basic state policy." This is the first time that Chinese environmental policy has been specified in the law, representing the development direction of the future civilization.

Judicial interpretation procedures for criminal cases of environmental pollution issued by the Supreme Court, and the Supreme Procuratorate have been officially implemented. For environmental protection criminal acts revolving multiple charges, the judicial adjudication follows the principle of "punishing the heavier one". In November 2013, the Ministry of Environment Protection announced on its official website the supervising and handling results of environment pollution issues pressingly concerned by the masses, with more than 100 power plants, coal mines, and brick plants in Changzhi, Shanxi involved and subject to severe punishment.

Besides, China is also developing the economic policy for ecological protection, to bring ecological damage and loss from environmental pollution

Foreword

The Environment Protection Law (Revised Draft) was adopted by at the Eighth Session of the 12th NPC Standing Committee on April 24, 2014, and shall take effect on January 1, 2015. It is the first time to incorporate ecological redline for strict protection into the law.

into national economic accounting system, establish ecological compensation mechanism, and actively explore the establishment of genetic resources accessing and benefit sharing mechanism; ecological monitoring system has been built, ecosystem monitoring capacity lifted, quality assessment of ecological environment carried out, and emergency treatment system for major ecological damages established; investment in ecological protection and construction has been increased, and a rational and diversified investment mechanism has been built by making full use of market mechanism, to establish and improve the ecology audit system, and to ensure the rationality and ecological benefits of inputs and outputs, and the unity between economic benefit and social benefit; ecological protection publicity has been vigorously carried out and efforts to create a public opinion atmosphere to conserve natural resources and protect ecological has been made, to raise the consciousness of the people to protect

Ecological Civilization of Contemporary China

Jianggudi glacier, just like a glacier museum, stood at the source of Yangtze River in 1998, but already melted away completely.

the ecological environment; international exchange and cooperation have been vigorously carried out, and foreign capital, advanced technology and management experience have been introduced, to lift the technology and management level of China's ecological environment protection. And China is actively participating in related international conventions, fulfilling its international obligations to safeguard national environment and development interests.

"Moving towards a new era of ecological civilization and building a beautiful China is an important part to realize the Chinese dream of bringing about the great rejuvenation of the Chinese nation. China will follow the philosophy of respecting the nature, conforming to nature, and protecting the nature, implement the basic national policy of resource conservation and environmental protection, more consciously promote the green development, cyclic development, low-carbon development, and blend the construction of ecological civilization into the all aspects and the whole process of economic

Foreword

Chinese people unite their wills in the construction of ecological civilization.

construction, political construction, cultural construction, and social construction, to form the spatial pattern, industrial structure, mode of production, lifestyle of saving resources and protecting environment, leaving the future generations with a production and living environment boasting blue sky, green land and clear water."

Blue sky and white clouds are the basis of people's livelihood and the common aspiration of the people. Good ecological environment is productivity, potential for further development, the core competitiveness and the cultural values. Pollution treatment is a protracted war and a tough fight benefiting the future generations. After declaring war against the pollution, the concerted efforts and unremitting work of all Chinese are bound to maintain the blue sky, white clouds, clear water and green mountains forever. Beautiful China will be our eternal homeland.

Ecological Civilization and Chinese Dream

Continuation, development and progress of human civilization are doomed to generate ecological civilization.

Today in China, the establishment of an ecological civilization where man and nature develop in harmony relates to people's wellbeing.

In November 2013, when the headlines of world's media were dominated by China's Third Plenary Session of 18th Central Committee of CPC, we could better understand the former media mogul Rupert Murdoch's sigh, "For the newspaper, the quickest way to be the best-selling is to put news on China on the front page."

And "accelerating the construction of ecological civilization system" is the biggest headline in China.

This is a social ideal, both noble and urgent. As General Secretary Xi Jinping has put it in his speech on July 20, 2013, "Moving towards a new era of ecological civilization, and building a beautiful China are the important parts to realize the Chinese dream of bringing about the great rejuvenation of the Chinese nation."

Reflections on the history of human development let us realize that although degree of civilization is increasingly high, the pace of civilization is mostly at the expense of natural environment. With the development of material civilization, more and more people start understanding the hazard of exchanging environment and ecology for economic prosperity.

Verdant jungle and mountains, crystal clear rivers and lakes, boundless azure sky, and fresh and pleasant air. For every life, the blue sky, green land, and clear water, are the most basic ecological civilization, and also an integral part of the Chinese dream.

In the face of resource constraint, serious environmental pollution, ecosystem degradation, respecting nature, conforming to nature, and conserving nature become the rational choice based on the ecological crisis

Rape flowers in spring in the eco-friendly Wuzhen Town.

and reflections on the traditional concept of development. And to make this rational choice, Chinese need to conduct a thorough reflection on transformation of the existing mode of production, living style and ideology, etc. If agricultural civilization is the so-called "yellow civilization", industrial civilization "black civilization", then ecological civilization is the "green civilization." Good ecological environment is the lifeline for "green rising". Construction of the ecological civilization boasting blue sky, green land, and clear water become the common expectation of all Chinese citizens.

Ecological civilization has never been an empty concept. It is the real life of everyone.

Only by respecting nature, will we be willing to follow the natural rules and revere nature; only by conforming to nature, will we more effectively protect and explore the nature.

Let polluted rivers be suffused with clear waves again; let grassland suffering desertification be covered by green grass; let poisoned land send off fragrance of clay again; let deforested mountains be surrounded by singing warblers.

Protecting nature, maintaining harmony between man and nature, as well as the benign interaction will create a more beautiful future!

Respecting Nature: Story of a Black-tailed Gull Called "Tianci" (Gift of Nature)

Weihai, Shandong. Hailv Island.

Over the boundless blue sea, tens of thousands of seabirds are soaring; on the solitary and steep cliff, a newborn black-tailed gull called "Tianci", enjoying the rising sun, is looking into the distance; in the beautiful purple flower clusters, its brothers are clumsily toddling...

This gull came into the view of the international film market in 2010.

At the Fourth Cologne International Film Festival, the feature film *My Garden of Eden*, on the background of an island of Weihai City, recording the

The photo excerpted from the film *My Garden of Eden*.

whole process of this bird hatching, learning to fly, and growing to an adult bird, won the best audience's choice award at the festival, becoming the biggest dark horse.

Known as the "China's first ecological feature film", it was filmed on Hailv Island. The endless sea, vast sky, and "Tianci" soaring between the sea and the sky, colorful and magnificent mountain and sea scenery in the film are a debut and the miniature of the natural ecosystem of Weihai.

The beauty of this film is that it didn't mean to arouse people's enthusiasm by making use of the common fact in the nature, but captured, without disturbing the normal life of birds, the great beauty that has never been reported. The flock of dancing, flying birds flew as beautiful elves on filmmakers' camera, and also flew into their heart.

At first, those birds were hostile to the crew, shitting on them and "bullying" them. They always landed on the camera, fighting for food with other birds. As the shooting progressed, the crew developed an increasingly strong relationship with the birds. They bought them fish, adopted young orphan birds, and released them after they were strong enough to fly.

Hailv Island is a treasure of natural creatures. The island is blessed with not only innumerable birds, but also various kinds of wild fauna and flora, like Venus fly trap, wild cabbage, edible amaranth, Shan Mu Zha, and wild garlic. In correspondence with the name of the film, that is the shooting location of *My Garden of Eden*.

After 7 years of zero-distance blending, in addition to finishing the film, the crew also had many new experiences and findings, completely new understanding and sense of the Mother Nature, particularly the respect and veneration for living creatures and the nature.

In the sunglow, the bird flock started again chasing and flying towards the blue sky.

Ecological Civilization of Contemporary China

Egrets, 60-cm long, have only white feathers.

Mountains and water, grass and trees, and men and birds, showed such a beautiful ecological picture.

Links

- Black-tailed gulls are the nationally protected terrestrial wild animals which are beneficial or of economic and scientific research values.

- Egrets are the second-class protected animals in China. In particular, the Chinese egrets were listed into the endangered species by the IUCN. According to statistics, the number of Chinese egrets is currently about 2500, with Hailv Island sharing more than 1000. They are as rare as the giant panda.

- In May, 2011, the crew of *My Garden of Eden* were rated as the second batch of moral models in Weihai by the citizens. As one of the important scenery spots of Weihai, Hailv Island used to have some inns on it for developing the tourism industry and a dam in the

Black-tail gulls, 47-cm long, are one of the largest gull species.

sea near it for attracting and carrying more birds. But later on, all the inns and the dam were demolished.

- From April 9 to 11, 2012, the 14th Congress of Party Representatives of Weihai established the strategy of "founding the ecological city, prioritizing the environment and realizing the green development." Environmental protection was made the primary element and precondition for developing the economy and society, and the "one-vote veto" was adopted.

The fresh sea breeze blows, several white clouds embellish the blue sky, and flocks of birds dance in the sun. This is the best gift of Mother Nature. Every Chinese deserves it. Respect the nature, and then you will enjoy gifts of the nature.

Conforming to Nature: Explore the "Chinese-style Environmental Protection"

The words "environmental protection" has never been as complicated and confusing as it is today.

What exactly do they relate to, politics or survival, a person, a city, a country or the whole world? What should the Chinese-style environmental protection be?

In a way, there is no ready-made answer, no model to imitate, or even no clear direction for now. At present, China must conform to the nature, grope forward on the rough road, to find a way out for environment protection.

Liao Xiaoyi, known as the representative of "Chinese-style environmental protection" on the rough road, gave up the chance to get a U.S. green card 10 years ago, she returned to China to establish the environmental protection agency - the "Global Village". Moving from the bustling city to the remote village, she earnestly advocates environmental protection, and explores another possibility of modern life. During the former 10 years, she spread the green concept, shot more than 100 TV and movies around the world, and won environmental protection awards in many countries; over the later 10 years, she was doing painstaking research on combining environmental protection with Chinese traditional philosophy, to explore the practical model for Chinese-style environmental protection.

Only by obeying laws of nature can we get protected by nature. Human is like a baby that looks very naughty, but can not leave the mother. If water and air are polluted, how will people survive? A yawn or a stretch of nature can be a destructive "natural disaster" for human, so the main ideas of environmental awareness and ecological ethic should be transformed from the vicious circle of anthropocentrism.

As the environment adviser of the 2008 Olympic Bid Committee, Liao Xiaoyi once wrote a song by the name of *Commitment*—

> When the bell of the new century has just struck, let's make a green commitment. For a healthy planet, for an eternal Olympic! We commit to get back the blue sky; commit to make every river clear; commit to plant the seed of hope; commit to live a green life. This commitment is a new trust; it is the statement of love; it is intuitive knowledge of kindness; it is choice of life.

According to ancient Chinese tradition, people attach great importance to be harmony with nature, to maintain health and attain mental tranquility. People never come into this world to consume materials. We consume materials just to survive. Therefore to the ancient Chinese, , desires for materials should be limited, and man should pay more attention to nourishing the spirit and inner world, which is the essence of ancient Chinese ecological civilization.

Obviously, Liao Xiaoyi expects every Chinese and every man on earth to follow laws of nature, to inherit the essence of traditional ecological civilization, and make the green commitment.

Based on reflection on that exploration, Liao Xiaoyi tries to set the following regulations for herself:

Take shower once a week, except in summer, to save water. Taking shower is not a routine of a day, but a rite of a week. Try not to use washing machine. It is good to wash clothes by hand and the laundry water can be used for flushing toilet. Never play golf, because the lawn needs to be maintained by large amount of agricultural chemical, which will seriously pollute the underground water.

Never buy or accept the cashmere products, because goats eating gross root influence the balance of ecology of prairie. Never buy cars. Sometimes take

Ecological Civilization of Contemporary China

Liao Xiaoyi is committed to explore the Chinese-style Environmental Protection.

taxi, but mostly take the subway or bus. Never buy the air-conditioner for her house. Sweating is detoxing and good for health, and the electrical fan is good enough. Never do sauna, because it is a big waste of energy. She is afraid to owe the nature too much. Never go to artificial ski resort, for the snow is made of the precious underground water, the source of drinking water. Spend some time in learning ancient stringed music instrument, Taijiquan, reading poems and doing meditation.

In the opinion of Liao Xiaoyi, green life is a green fashion. The core of green life lies in the moderate consumption, shrinking the ecological footprint to the largest extent and reducing the cost of environment.

Returning to traditional philosophy and nourishing the spiritual space require us to "respect nature and cherish materials". It can be explained by acronym "PHD", namely, "primativity", meaning that natural things are the best (original nature), "holistic", meaning that all things on earth are associated (association), and "primitivity", meaning that differences are normal (diversity).

Ecological Civilization and Chinese Dream

To "respect nature and cherish materials" doesn't mean to stop taking from the nature, but to follow laws of nature and consume materials in a moderate way.

For Liao Xiaoyi, "Global Village" and other NGOs, converting the green commitment from individuals' life comprehension, actions and opinions to the exploration of model of Chinese-style environmental protection is very important and of practical significance.

After all, it is impossible for only one person or an environmental protection NGO to change the world, no matter how powerful they are. We should convert "conquering nature" to "conforming to nature", to realize the harmony between human and nature. Getting rid of the gaudy appearance, Chinese-style environment will win the wisdom and health.

Links

- Liao Xiaoyi established the Beijing Global Village Environmental Education Center in 1996. She used to promote the environmental

Gyaring Lake at the source of Yellow River is "shrinking".

protection philosophy through televised documentary, acting as the independent producer of program Environmental Moment on CCTV, and the environment adviser of the 29th Olympic Committee. She won the Sophie Prize in 2000, the "Banksia International Environmental Prize" in Australia in 2001, and the "Clinton Global Citizen Awards" in 2008.

- One month after the 2008 Wenchuan earthquake, Liao Xiaoyi and the Global Village team went to Sichuan to investigate the situation of the disaster. In the razed Daping Village, Tongji County, Pengzhou City, they were surprised to find an old wooden-structural house safe and sound. Inspired by setting the country buildings on the road to ecological civilization during the post-disaster reconstruction, Liao Xiaoyi had the idea of constructing a "Harmonious Homeland". Today, the "Harmonious Homeland", as an experimental model of eco-houses, has been built to a large-scaled ecological residence. And it is the result of reflection of Liao Xiaoyi on Chinese-style environmental protection.

- On January 8, 2013, to explore the new natural ecology and rural environmental protection road of developing economy while protecting environment and protecting environment while developing economy, the general office of Ministry of Environmental Protection issued the circular *2013 National Natural Ecosystem and Rural Environmental Protection Work Points*. The circular specified the important instructions on exploration of inter-regional ecological civilization construction, exploration of the industrial working mechanism to promote ecological civilization, and establishment of industrial ecological civilization demonstration base.

Former UN Undersecretary Maurice Strong takes part in the activity of the "Global Village".

Economy, thrift, dedication, respecting nature and cherishing materials are all environmental protection. Following laws of nature is environmental protection, which gives Chinese-style environmental protection courage and confidence. Energy saving and low carbon culture, starting from the source, are more effective than changing the confirmed habits hard to be got rid of. This is the significance of exploration for Chinese-style environmental protection.

Protecting Nature: Chinese Environmental Protection Dream

When talking about protecting nature in China, we must mention the Alxa SEE Ecological Association. Since the name sounds a little "regionalized" and even "frontier" and not magnificent, it seems that it's just a little "folk group" without influence.

Alxa means "colorful" in Mongol: Among those colors, green refers to forest on Helan Mountain, yellow Badain Jaran Desert, blue the desert lake, red weathered sandstone, white the salt lake, purple agate, golden the populus euphratica forest in autumn, and black the sandstorm! The sandstorm arises from Alxa and hits Beijing for many times, which is a big annoyance for people in Beijing, including the "rich" entrepreneurs.

Society of Entrepreneurs & Ecology (SEE) participates in sand storm control.

In February 2004, initiated by Liu Xiaoguang, general manager of Beijing Capital Group, a number of entrepreneurs in the dust shrouded Beijing decided to set up an environmental organization, to control the desertification in the source - Alxa, Inner Mongolia The environmental organization is the Chinese Alxa SEE Ecological Association funded by nearly a hundred well-known Chinese entrepreneurs ("SEE" in short) in June, 2004.

As the initiators, the nearly one hundred entrepreneurs promised: Invest RMB 100,000 annually per person for 10 years in a row, to slow down sandstorms in Alxa, protect the environment in China, promote harmony between man and nature, between man and society, and among people. Facts have proved that during the development of SEE, not only the number of members, but also the donated money sees increase every year. SEE stands for a new beginning for Chinese entrepreneurs to collectively and consciously protect the environment and bear the social responsibility.

For the desertification control in Alxa, at first, many entrepreneurs just simply assume that increasing oasis and planting trees are the solutions. But later they learned that extracting groundwater to water trees is not a good way, for even if the tree grows, other parts of desert ecosystem would have been destroyed. Desert itself is in balance, and the fundamental way to control desertification lies in the protection, especially to prevent excessive human activity accelerating its degradation. Sandstorm already exists for more than 2 million years, and the area that can be controlled by human is just 60,000-70,000 square kilometers. Thus, ecological management should target at a natural conservation with limited objective. So on June 5, 2004, initiators of Alxa SEE Ecological Association issued the *Alxa Declaration* by Moon Lake of Tengger Desert. In this declaration, all of them ask questions and answer themselves—

> Why do entrepreneurs like us come to Alxa from all over the country? Why do we set up the "Alxa SEE Ecological Association" to participate in the cause of China's sandstorm control? Because we

Populus diversifolia—oasis in Ejina, Alxa.

have hopes and dreams, hoping that Chinese economy will be more prosperous, people will be more affluent, relations among people will be more harmonious, and land of China will carry beautiful mountains and clear water, full of vigor, for ever. And we hope that people around the world live together on the beautiful earth, and we dream of a place where everybody has the opportunity to realize their aspiration.

The *Alxa Declaration* is a declaration and solemn commitment of initiators of Alxa SEE Ecological Association. But this hope and dreams should also be declaration and commitment of all Chinese enterprises and all Chinese people to protect the nature.

After nine years of hard work, when people come back to Alxa, large pieces of beautiful oasis present before them.

Links

- Alxa SEE Ecological Association is a membership-based NGO, pursuing the non-profit principle. "SEE" is an acronym of three words, society, entrepreneur, and ecology. With the vision to promote the sustainable development of man and nature, it follows values of achieving unity among ecological, economic and social benefits. The aim of SEE is to start from desertification control in Alxa region to solve the desertification through comprehensive development of communities, meanwhile, to impel Chinese entrepreneurs to assume more responsibility for the environment and society, and to promote the environmental protection and sustainable development construction of companies.

- On April 6, 2010, Alxa League introduced policies to encourage the society to contribute to the afforestation and desertification control.

Golden populus diversifolia forest in Ejina.

As specified in the policy, in the future those who participate in the afforestation and desertification control in the surrounding regions of key towns of Alxa, making forest of 500 *mu* (1 *mu* ≈666.67 m^2) or larger, accepted by the forestry department, will gain a one-time subsidy of RMB 60–100 (not less than RMB 60 for shrubbery, and not less than RMB 100 for high-forest) per *mu* from the government.

- With efforts from various parties, the rate of desertification in Alxa has decreased from 1000 square kilometers/year in 1990s to the current 353 square kilometers/year.

To protect nature, and promote harmony between man and nature, between man and society, and among people, in addition to protecting air, water, and soil, we must change the lifestyles and ideas of excessive consumption. Protect the nature, let it rehabilitate, and trust its power. The picture of "As wind lowers grass livestock cannot hide" in Alxa will not just be a dream.

Blue Sky

Compared with water and food, air is human's largest consumable.

Every human being needs to consume 12kg of fresh air every day. However, the current China has not fundamentally changed its extensive industrial model featured by high consumption, high pollution, low output, and low efficiency. With a large number of pollutants discharged, as well as the progress of ongoing industrialization and urbanization, China is facing unprecedented environmental pressures.

In 2013, the national average number of smog day was 4.7, 2.3 days more than previous years on a year-on-year basis, making 2013 the year with highest number of smog days over the past 52 years. PM2.5 has become the normality we have to face.

Ecological Civilization of Contemporary China

Blue sky and white clouds after raining over Wutai Mountain in May 2013.

Now when it is a fine day with blue sky and white clouds, people will get excited like receiving a year-end bonus. Originally, enjoying a fine day is our right, but now it has become a luxury and surprise. People 30 years ago attached great importance to their living standards, but now we highlight ecological environment; people at that time just sought for adequate or ample food and clothing, but now we call for environmental protection.

Why? Just imagine: if the wealth we accumulated over our lifetimes is just finally paid for medical expenses, is it worthwhile? Doesn't such a serious situation deserve our deep reflection?

We can not choose the surrounding air, but we can not ignore it. On the contrary, we need to do something and give people the confidence and strength.

After all, life still needs to move on. A more blue sky or darker sky overhead totally relies on what we choose and what we do.

Why do we always sorrow with full tears in our eyes? Because we love it so deeply.

We feel the same pain as its pain, suffer the same bitter as its bitter, delight with its clear, and long for the same yearn as its yearn.

White clouds in the blue sky are just like the blossoming magnolias, which dispel fog and bring us shine.

Let us select the blue sky. Let our freely jumping hearts be touched by the free, fresh and magnolia-like blue sky.

Let's leave a touch of blue for ourselves and reserve a blue sky for our offspring.

PM2.5: A Reformation Caused by a Cough

Pan Shiyi, a well-known real estate tycoon and a network big V (verified weibo users who have more than 500,000 followers), has been in a blue mood recently, because his wife (named Zhang Xin) has frequently suffered from cough. She coughed heavily in the heavily polluted days; got relief in better days, and stopped cough in unpolluted days. This situation has been lasting for a long time. Why did it happen?

His warm-hearted friends reminded him that it may be caused by PM2.5.

As for the attention to the issue of PM2.5, Pan Shiyi said it was purely accidental. His wife coughed badly as soon as she returned from abroad. In virtue of a friend's reminder, he then learned that there is a kind of small particle in the

Pan Shiyi (right) President of SOHO China and Zhang Xin, CEO of the group.

Blue Sky

Blue sky is seen in Beijing after several days of haze.

air, which can enter directly into human's alveoli. Since then he began to pay attention to PM2.5.

Fortunately there was an air quality monitor set up in the U.S. Embassy located at the 3rd Ring Road East, Chaoyang District, Beijing, which was originally used to provide air quality index for internal personnel in the U.S. Embassy. The warnings of "heavily polluted" frequently issued by this monitor were posted on weibo and the Internet by many enthusiastic netizens. Pan Shiyi thought there is no reason not to spread these warnings to more people. So he began his broadcast on his verified weibo from October 22, 2011. Later on, He spent hundreds of thousands to buy a set of air monitoring equipment to continue to broadcast PM2.5.

At that time, China had no official PM2.5 standards and monitoring. We all know the things later: the government also began to issue PM2.5 data, which, however, were so different from those issued by the U.S. Embassy, making the public doubt.

Ecological Civilization of Contemporary China

Pan said that in fact it does not need the equipment, because his wife' throat can measure Beijing's environmental pollution: if the environment is not polluted, his wife will not cough in the morning; if the air pollution is slight, she will slightly cough; and if the air pollution is serious, she will heavily cough in the morning.

His words are not unfounded. The data show: Beijing's air pollution is most serious among China's five largest central cities; Beijing ranked 42 from the bottom among global 1085 metropolises with this regard in 2011. Many environmental experts have suggested that PM2.5 does affect the image of the capital, but at the moment PM2.5 causing health damages to people living there deserves more attention.

It was just such a seemingly simple indicator that resulted in endless "verbal grenades" among netizens, even once upgraded as verbal confrontation between China and United States. For this reason, Pan Shiyi disclosed on his weibo that he was once invited to "drink tea" (a euphemism for getting investigation or instruction) by relevant departments.

Of course, this story also had a "comforting" ending. A few months later, the new air quality standards were released. People who invited Pan to "drink tea" later became his friends.

Links

- PM2.5 refers to the particles at aerodynamic equivalent diameter of 2.5 μm or less in the ambient air, is also called fine particles, which can suspend in the air for a longer time. The higher its content (concentration) is in the air, the more serious the air pollution is.

- Recently, a number of medical studies have shown that particles at diameters of 2.5 μm or less caused most intense damages on the human body. Because parts of particles at diameters of 10 μm to 2.5

μm will partly be exhaled in respiration, the others will be adsorbed in the bronchial and lung tissues; while particles at diameters of smaller than 2.5 μm (microns) will directly enter into the gas exchange region (alveoli) of human's lungs, and retained in the deep of lungs. In addition to their stimulation and sensitization to human's respiratory system, they also may invade human's lungs as carriers with bacteria as microbes, viruses and cancerogens, which will seriously harm human's health.

- On February 29, 2012, Premier Wen Jiabao chaired a State Council executive meeting, agreed to release the newly revised Ambient Air Quality Standards. The new standards incorporated monitoring indicators for fine particulate matters, i.e. PM2.5, and ozone's 8-hours concentration limit.

- China began to implement the new Ambient Air Quality Standards

The construction site for reconstruction of the "village in the city" of Beijing.

Ecological Civilization of Contemporary China

from January 1, 2013 and conduct real-time monitoring and release data of 6 basic items including PM2.5, as well as AQI index (ambient air quality index).

- In early June 2013, Pan Shiyi was granted the Special Jury Award at the fifth SEE-TNC Ecological Award due to his promotion to issuance of China's PM2.5 air quality standards. The reason for this award is in that in the PM2.5 civil environmental movement from the end of 2011 to now, Pan Shiyi, who had over 10 million Weibo followers, continuously broadcast Beijing's air quality, which promoted and accelerated the introduction of China's PM2.5 air quality standards.

Today's Chinese people collectively began to be fascinated with days with blue sky and white clouds. On the one hand, people's focuses on PM2.5 are closely related to the improvement of environmental awareness, sense of justice

Haze control slogan raised in front of an under-construction residential building in Huai'an City, Jiangsu Province.

in people's hearts, as well as concerns for our living environment. On the other hand, as the words that Bai Yansong, a well-known television host, quoted from one of his German friends that "for current China, the air seems to be more important than democracy, freedom, and GDP", which shall be more worthy of our thinking.

RMB 5 Trillion for 5 Years:
The Capital City Surrounded by "Haze"

After October 2011, many Beijing citizens who longed for spending "the most beautiful season of the year" were disappointed, because the continuous haze weather led people feel cold and breathed hard. Unfortunately, the best season of the year became the "worst" one.

At the same time, the topics on air pollution quickly spread among the people.

There is a piece of irony words saying that, the furthest distance in the world, is not between life and death, but when I stand in front of you, yet you can not see me.

"I can neither breathe freely, nor always see the blue sky; work environment like this seems not for a white-collar but a prisoner. When I stand before the windows, I can hardly see the office building less than 100 meters in front of me; the whole city is filled with unhealthy air composed of yellow-brown dusts, which fly under our eyes into our contact lenses, let us be demoralized, and weaken our health; that's the life of current white-collars.", this is a piece of expression written by a white-collar to describe his or her current situation, which has been well recognized by many netizens.

Wearing neat suits, sitting in the office building, working hard at the table and overlooking by the windows once in a while was once the life style of white collars in CBD. But now, the gloomy haze makes such elegance extraordinarily depressed. A colleague based in Beijing chose to work back Shenzhen on the eve of the Spring Festival. He was pleased to find that his old illness-rhinitis which

Post-haze time. A nearly completed building in Beijing.

has tortured him over years in Beijing miraculously disappeared in just a few days there.

The colleague escaping from the capital city surrounded by "haze" is very happy for his choice. Despite air quality in Shenzhen is still not perfect, but it is much better than that in Beijing.

In Northern China, contents of the most harmful toxic pollutants in air had reached 40 times of the recommended levels by the World Health Organization (WHO). More importantly, by releasing and comparing the PM2.5 data, people can easily relate these data to the air they are breathing, therefore, the air pollution which has not attracted the public's enough attention has become a major source of public angers.

Atmospheric pollution problems also emerged in developed countries in their course of industrialization over hundreds of years. In the world's eight pollution incidents in the history, including the London Smog Disasters, Los Angeles Photochemical Smog Episode, Belgian Maas Valley Smog Event, USA Donora Smog Event, five events resulted from the hazy weather caused by industrial smoke and automobile exhaust emissions, etc.; the other three events were also caused by the regional rapid industrialization.

Public campaigns are still escalating, and hazy weather still rarely goes worse.

On September 23, 2013, Beijing's 8 bureaus and commissions including Beijing Municipal Environmental Protection Bureau, Commission of Transport, Commission of Development and Reform, Commission of housing and Unban-rural Development, etc. jointly held the "Beijing Clean Air Action Plan" press conference, where Fang Li, the deputy director of the Beijing Municipal Environmental Protection Bureau and the press spokesman, expressed that, Beijing would vigorously control air pollution, and was expected to invest RMB 5 trillion in the next five years. The government will invest about RMB 200 billion to RMB 300 billion; the whole society will invest RMB several trillions.

The plan is mainly designed to control PM2.5, which specifies that the annual PM2.5 average concentration in Beijing's air in 2017 will decrease by more than 25% over that in 2012, down to about 60 mcg/m3.

Data show that, in the past 10 years, haze days in 2012 reached maximum to 124 days. In the past 30 years, those days in 1980 reached maximum to 135 days. Since the establishment of Beijing Meteorological Station, those days in 1951 reached maximum to 261 days.

Chinese Academy of Social Sciences and China Meteorological Administration jointly issued the *Green Book of Climate Change*, which also pointed out that, in 2013, the national average number of fog and haze days were 4.7 days, 2.3 days more than normal ones, which is the longest since 1952.

Blue Sky

Old building renewal in Beijing under the haze weather.

According to statistics of the National Development and Reform Commission, early in 2013, China has undergone a wide range of continuous fog and haze weather; the affected areas included the North China Plain, the area south of Yellow River and north of Huaihe River, Yangtze River and Huaihe River, Jianghan Plain, regions in the south of the Yangtze River, and north of Southern China; the affected area accounted for about 1/4 of China's national territorial area; and the affected population amounted to about 600 million people.

Links

- In December 2012, the Ministry of Environmental Protection, the National Development and Reform Commission and the Ministry of Finance jointly issued the *12th Five-Year Plan on Air Pollution Prevention and Control in Key Areas*, which requires to strictly conduct environmental access, to create the "forced transmission mechanism" to promote economic development through environment optimization.

Ecological Civilization of Contemporary China

- In June 2013, the Supreme People's Court and the Supreme People's Procuratorate issued the *Interpretations for Several Issues on the Applicable Laws for Handling Criminal Cases on Environmental Pollution*.

- On September 12, 2013, the State Council issued the Air Pollution Prevention Action Plan. The Plan proposes to improve the overall air quality in the country with efforts of five years and substantially reduce the heavily polluted days, and markedly better the regional air quality in Beijing-Tianjin-Hebei, the Yangtze Delta and the Pearl River Delta regions. Its specific objectives include that: by 2017, the inhalable particulate matter concentration ratios in cities at prefecture level or above decrease by more than 10% over those in 2012 with good days gradually increased; concentration of fine particulate matters in Beijing-Tianjin-Hebei, the Yangtze Delta and the Pearl River Delta regions fall by about 25%, 20%, 15% respectively.

Tiananmen Square in haze.

- On the morning of January 18, 2014, the Beijing Air Pollution Control Regulations (draft) was submitted to the Beijing Municipal People's Congress for deliberation, and it was the first time that PM2.5 reduction had been included in the legislation. Wang Anshun, the Mayor of Beijing, publicly expressed his determination to control haze, and he vowed with his life to achieve the objective for air improvement by 2017. Wang Anshun said that, in order to realize the blue sky, clear water, and green land by 2017, the investment of RMB trillions is worthy.

On the morning of January 22, 2014, the Beijing Air Pollution Control Regulations upon three deliberations was submitted and adopted in the Second Session of the Fourteenth Beijing Municipal People's Congress, and officially implemented from March 1, 2014. The voting results included 659 votes in favor, 23 votes against and 14 abstentions. The Regulations definitely propose to take "reducing PM2.5 concentration" as the key point of air pollution control, and control pollutant emissions in whole process from source to the end, to accelerate the reduction of total pollutant emissions.

Since the implementation of the Beijing Air Pollution Control Regulations, Beijing Municipal Environmental Monitoring Corps opened their first ticket on March 16, 2014. They delivered the written decision of administrative penalty to Beijing Hongxianghong Heating Power Co. Ltd., the first unit investigated due to breaking laws, to order this company to correct their violations within three days, and impose a fine of RMB 100,000. This company didn't add adequate alkaline substances in the desulphurization section of spray precipitation in smoke emission, which caused that the sulfur dioxide in the smoke they discharged seriously exceeded the standard.

Life still needs to move on. A more blue sky or darker sky overhead totally relies on what we choose and what we do.

This "first haze ticket" carried the public's expectation and hope, which was a direct declaration of war against pollution. Although controversial, it is of both practical and landmark significance.

We all expect that Beijing will be a city with pleasant and refreshing air in 5 years later.

Urban Ventilated Corridor: Hangzhou's Actions of "Introducing Wind and Dispersing Haze"

Urban ventilated corridor has been a very fashionable new term on environmental protection.

Figuratively speaking, the urban ventilated corridor is a corridor that runs through the city and lets the wind freely flow in the city, which, however, does not require demolition and construction on a large scale.

In summer, we can obviously feel the alleyway wind while standing between two high buildings. Alley between two buildings is called the "ventilated corridor". Therefore, the ventilated corridor is not a fixed object, but just a "big alley" running through the city without obstructions, through which the wind can quickly flow in and out.

The urban ventilated corridor may be understood as a visual reconstruction of the city, which is designed to solve the problems of space and survival dignity. On the one hand, a green corridor of the city needs to be built, namely, a green belt with modest scale needs to be planned along main roads, highways, railways, rivers, lakes, etc. around the city; on the other hand, the urban planning shall be reasonably considered, such as adjustments to directions of streets, roads as well as building distribution, so as to set apart a corridor for wind's free passing.

In a sense, the urban ventilated corridor is an idea of urban structural transformation and improvement of urban functions, as well as a new environmental strategy based on local conditions, which can effectively introduce the wind from outskirts into the main city, and blow the haze and other air pollutants in urban air away. If a high-rise is just planned in the ventilated

Ecological Civilization of Contemporary China

In June, 2014, Houbeitun Village in Taiyuan, Shanxi Province, was included in the reconstruction project as a major village in the city to be dismantled in 2014.

corridor, the urban planning department shall consider its feasibility, to cancel this planning or let it change the direction to not hinder the ventilated corridor. For example, Munich, Germany built five ventilated corridors in its urban planning, so as to enable the seasonal hot wind to pass through the city and bring away the city's dirty air, which have gotten good effect.

The unique approaches on "introducing wind and dispersing haze" have been gradually highlighted in China's urban planning with some staged attempts in some regions. The Century Avenue in Shanghai and ecological ventilated corridor in Nanjing and others are respectively built and playing their roles.

Xu Kuangdi, an academician of the Chinese Academy of Engineering, said that, in the planning for construction of Shanghai Pudong New Area, "ventilated corridors" were deliberately left out to enable the wind to freely pass through. "Shanghai blows southeast wind in summer and mostly northwest wind in autumn, so a century avenue of 250 meters wide was planned and built

Landscape along the Century Avenue of Shanghai.

in Pudong. It was not to go in for pomp but enable the wind blow in", thus, the 250-meter-wide Shanghai Century Avenue became a "ventilated corridor".

Nanjing planned the "ventilated corridors" relying mountains, forests, and valleys. Jiang Ling, a member of expert group for editing and revision of Nanjing's overall planning and professor of Nanjing Tech University, said that, Nanjing reserved 6 ecological ventilated corridors; some were three or four km wide, and some are one or two kilometers wide, and their lengths basically had tens of kilometers, so as to form broad and large "vents" for the city through taking advantage of mountains, woodlands, valley waterways, wetlands, greenbelts and other natural conditions, as well as means such as higher limit for construction. Currently, most of the eco-ventilated corridors have been basically formed and playing their roles.

Coincidentally, Hangzhou City also suffered from the "haze" in a season which shall have been terrific and suitable for going out for travel, and the whole city was covered by dusty sky. In November 2013, Hangzhou Planning Bureau, and Environmental Protection Bureau, after careful consideration, decided to start the research and attempts for construction of urban ventilated corridors to control the haze.

A large number of dense high-rises artificially separate the urban ecosystem over time, which makes the urban atmosphere difficult for virtuous cycle; the wind can not come into the city and the haze can not get out, the city becomes a suffocating polluted "besieged city". As a city surrounded by mountains in three directions, Hangzhou's building of ventilated corridors can form a ventilation opening in local areas in the city, allowing the wind from outskirts to blow into the main city to increase air flow in the city, which can effectively mitigate the haze in the city.

Hong Shengmao, an engineer of Hangzhou Environmental Monitoring Center, said that, the number of days with good air quality in Hangzhou in 2013 totaled 198, with the good rate of 65.9%.

According to the research of Hangzhou National Reference Climatological Station, the increase trend of hazy weather in Hangzhou has been controlled effectively. Before 2005, the frequency and the number of days of hazy weather trended to keep stable and decreased slightly; and the atmospheric pollution control and urban eco-environmental improvement in Hangzhou had achieved initial success. Xiang Zhen, Deputy Director of Pollution Prevention Department of Hangzhou Municipal Environmental Protection Bureau, introduced that the Bureau was researching and developing emergency response plans of heavily hazy weather, which may include temporary restriction or suspension of production of heavy pollutant discharge enterprises.

Air flues are like the "main and collateral channels" of a city; if the air flues are blocked up, there will be something wrong with the city. Taking a broad view on most of cities, what they lack of at present is such prospective planning. New

The sea breeze is blowing away the haze in Xiamen, Fujian Province.

environmental protection policy opens a new "wind" path, "catching the wind and dispelling the heavy fog". Let's wait and see what will happen.

Thorough Treatment of Pollution in Lanzhou

When the flying sparrows are wearing dusty "armors", the winter of Lanzhou is coming quietly. Only by standing on the top of Lanshan Mountain, can we know it's so hazy to live under grey fly ashes, which bring the whole Lanzhou into a dusty world.

Let's have a look at the "grey records" of Lanzhou in recent years: in 2005, there were 238 days with Grade I and II air quality; in 2006, there were 205 days with Grade I and II air quality; in 2007, there were 271 days with Grade I

Zhongshan Bridge, the "First Bridge of Yellow River" in Lanzhou, Gansu province, is buried in the dust.

and II air quality; in 2008, there were 268 days with Grade I and II air quality; in 2009, urban atmospheric pollution rebounded, and there were 236 days with good air quality. However, in early 2010, there was abnormal weather that the concentration of SO2 exceeded the grade III standard of national air quality in Lanzhou City.

In this morning, Jing Jun, Head of Yanjiaping Sub-district Administration, came to the platform on the top of office building to observe big and small "pollution sources" above the city as usual, and then took out his carry-on notebook and wrote down the conditions. After recording these "pollution sources" in details, he assigned staffs to survey relevant conditions and give some advice. He has done such work for more than one month.

The "pollution sources" become more and more instead of decreasing, which made citizens' complaints heard everywhere and strongly influenced the path of opening up of Lanzhou. Personnel from authorities from Municipal Party Committee and Municipal Government to sub-district offices all cannot turn a blind eye to it with indifference any more.

Links

- On November 23, 2010, according to the daily report on air quality of 86 key cities published by the Ministry of Environmental protection, the pollution index in Lanzhou was 353 in that day, while its air quality was heavy air pollution; Lanzhou became the city with the worst air quality nationwide. Since November 21, the pollution index in Lanzhou City had exceeded 301 and been heavy pollution for three days.
- According to the circular issued by Lanzhou Economic and Information Commission and Lanzhou Environmental Protection Bureau, all brick and tile enterprises would suspend production from November 25, 2010 to March 31, 2011.

In March, 2008, a worker is cleaning garbage dumps floating on the yellow water at the Yellow River Trash Barrier near the Xiaoxia Power Station in Shichuan Town, Gaolan County, Lanzhou province.

To chop off the "pollution sources" fundamentally, correctly master the current situation and causes of atmospheric pollution of Lanzhou City, and govern pollution with a focus on industry, fuel coal and non-point source, motor vehicle exhaust, reentrainment dust pollution, etc,. Lanzhou City issued six combinations of measures for addressing the symptoms and cause simultaneously—

1. Relocate heavily polluting enterprises first. Give different guidance to different categories of industrial enterprises, renovate them in a defined period, relocate a group of them, reform a group of them, and close a group of them, on the basis of actual conditions of regional such industrial enterprises.

2. Accelerate urban energy structure adjustment. Adjust energy structure, accelerate the implementation of "West-to-East Heat Transmission" project with pyroelectricity as heat source and the construction of heat supply network, and expand the area of centralized and contiguous supply of heat.

3. Vigorously improve the control and governance of motor vehicle exhaust pollution. Manage the resource control of "registration, test and recycle" of motor vehicles well and actively popularize and use new energy automobiles.

4. Strictly control fugitive dust pollution. Encircle and shelter all construction sites of buildings, demolition, the excavation of road, etc.; take dust suppression measures (e.g. hardening) for all bare ground.

5. Accelerate the building of urban-rural integration eco-network. Implement the 1-million-*mu* green ecological construction project.

6. Transform the industrial structure. Arrange regional environmental resources comprehensively, take over the industrial transfer and environment stress of old district of the city, and meanwhile, accelerate and promote the construction of the mountain-moving and road-building project.

In 2012, there were 270 days with good air quality in Lanzhou all year around with a good rate of 73.8%; and it is a year with most days in good weather in past five years. In the monthly report on air quality of some cities published by China National Environmental Monitoring Center in January 2013, Lanzhou City ranked the 36th among nationwide 74 newly identified key cities which published air quality information.

Through the thorough treatment of pollution, we have made some achievements. Nowadays, we have eliminated the "pollution sources", and welcome this eco-friendly city with endless vigor and colors—Lanzhou.

On April 16, 2013, *Measures of Implementing Law of Prevention and Control of Atmospheric Pollution in Lanzhou (revised draft)* was passed at the executive meeting of Lanzhou municipal government. The draft, a local

Ecological Civilization of Contemporary China

Blue and clear sky and fresh air in today's Lanzhou.

legislation, was designed to put forward stricter enterprise emission standards, implement big pollution control and emission reduction program and carry out the strictest environment supervision measures.

If pollution control is the embodiment of willpower, then we can say that treating both symptoms and root causes is the reflection of executive force.

A good therapy concerning pollution control will change pollution sources into benefit sources and everyone should take their responsibility for the pollution control in this city to protect our home.

Industrial Relocation: Capital or Shougang?

On July 13, 2001, Beijing won the bid to host the Olympic Games.

With celebration around the whole country, Rong, veteran worker of the First Steel Works (western suburb of Beijing, Shijing Mountain) of Shougang Group, seemed in a low mood because "Shougang is the enemy of the blue sky and Green Olympics". As most of the workers in Shougang, he felt pity that one era for Shougang was about to come to an end.

Beijingers have a complex feeling for Shougang. Shougang changed from the Shijing Mountain Steelworks, which was started in 1919. After new China was founded, Shougang became one of the few businesses that were titled with "Shoudu", which means capital and it was also one of the biggest industrial businesses in Beijing. In the heyday, profits and taxes from Shougang can even account for 1/4 of all the taxes in Beijing.

As the industrial business model, Shougang provided a lot of work opportunities. At the end of 2003, workers of Shougang amounted to 135,100, accounting for 1/6 of all workers in Beijing. And without any doubt, Shougang was the main source for fiscal taxation of Beijing. From 1979 to 2003, profits and taxes delivered to the state by Shougang amounted to RMB 35.8 billion.

From this, we can say that Shougang plays a pivotal role in the economic and social development in Beijing. However, on the other hand, Shougang was also the main pollution source for Beijing for several decades.

"There is a black cover in the sky of Beijing with Shijing Mountain in the center and Shougang is just under it". This is how Beijingers described the causal chain between Shougang-Shijing Mountain-Beijing in the earlier time.

Ecological Civilization of Contemporary China

Because "Green Olympics" is the key concept for Beijing to organize the Olympic Games, pollution like this is intolerable. Beijing Shougang Plant is only 17km away from Tian An Men, seriously influencing the air environment quality of Beijing. The worst is that it locates in the windward and upstream place, and the west wind will bring the smoke dust to every place of Beijing. Beijing fails in applying for holding Olympic Games at first time, environmental protection not meeting the standards is an important factor.

"Beijing or Shougang" is a nerve-wracking problem. In order to hold the green Olympics, relevant authorities made up their mind to move Shougang out of Beijing.

The relocation of Shougang gives blue sky and green land back to Beijing. The first shaft furnace stopped production, reducing emission of sulfur dioxide by 48 tons, organized dust by 100 tons and unorganized dust by 84 tons every year.

On December 21, 2010 in Shijingshan District of Beijing, several workers were having a rest on the grassland before the No. 3 furnace which had stopped production.

The No.2 code oven stopped production, reducing emission of sulfur dioxide by 0.24 tons, smoke by 4.05 tons, and dust by 189 tons. Smelting and hot rolling production lines in Shijing Mountain completely stopped production, which will realize the zero emission of dust, smoke and sulfur dioxide.

Shougang Group witnessed the change of time and the improvement of ideas. Aristotle, a philosopher from ancient Greek, had said: "People gather to a city for surviving, and reside in the city for better life." But, people have to form vigilance and think of urban functions, for the noise and traffic jams caused by excessively rapid development of big cities. Serried chimneys make blue sky and clean air become less and less. It is impossible for people to live better in such cities.

If we want the city to be more livable, we have no choice but to move the heavy chemical industries out of it. This is the pain we must suffer from because Shougang moving out of Beijing brings room to more cities to think about themselves.

Exhaust gas from a heating chimney in Beijing.

Ecological Civilization of Contemporary China

Green scene in the Beijing Botanical Garden in 2012.

Links

- From late 1980s to early 1990s, the annual steel output of Shougang Group jumped from 1 million tons to more than 8 million tons, ranking first in China. In early 1990s, within the 86 square kilometers in Shijingshan District, the worst dust emission per square kilometer every year got 34 tons, on average. According to statistics, the inhalable particle emission in Beijing in 2002 reached 80,000 tons, and Shougang Group accounted for 23%.

- In February 2005, the National Development and Reform Commission officially approved the production reduction, plant relocation, structure adjustment, and environmental governance of Shougang Group, and the establishment of a new Shougang Group in Caofeidian District, Tangshan, Hebei. The construction started in March 2007, with two 5500-cubic-meter shaft furnaces claiming No.3 in the world and NO.1 in China built. They were put into operation in 2010.

- On January 13, 2011, the government of Beijing passed the Shougang Industrial Area Transformation Plan, to transform the former site of Shougang to be a new urban district, positioned as the "post-industry cultural and creative industrial zone".

There is no doubt that, no matter it is the pollution treatment or the industrial structure adjustment, Shougang, developing during the relocation, made a big contribution. Now in Beijing, sky is bluer and water is clearer, not because the development stops or slows down, but the industry sees a new development at a higher level.

Joint Prevention and Control: "Pilot Pollution Control" Program of Beijing-Tianjin-Hebei

In the spring of 2013, the tough battle of suppressing the haze started.

As the report from Ministry of Environmental Protection showed, environmental protection equipment of more 70% of the 300 steel companies interviewed in Beijing-Tianjin-Hebei failed to meet the standard. Crude steel output in the four provinces around Beijing in 2012 reached 400 million tons. The sulfur dioxide, nitric oxides and dusts emitted by the high-energy-consuming, high-polluting and high-emitting steel industry are believed to be the "culprit" of air pollution.

The four provinces around Beijing, Liaoning, Shandong, Hebei, and Shanxi, distributing from east to west, share a common characteristic, big producer of steel and iron.

The battle has developed to be a political assignment, Hebei, with intensive heavy industry and serious structural pollution, has become the key region in the pilot program.

On January 12, Ministry of Environmental Protection released the urban air quality daily report, in which the most serious four cities were all in Hebei Province. Shijiazhuang, Handan and Baoding grabbed the top 3 places with pollution index of 500, followed by Tangshan, where pollution index was 458. According to monitoring data of the air quality issuing system of Hebei on January 12, all the prefecture-level cities suffered heavy pollution, except for Zhangjiakou and Chengde.

Blue Sky

Beijing is immersed in haze for which yellow warning of heavy pollution was issued.

According to regulations, the air pollution index is divided into six levels: Level I refers to pollution index of 0-50, an excellent air quality condition; The larger the index is, the worse the pollution is; When it exceeds 300, the pollution reaches Level VI, meaning a heavy pollution; And the highest pollution index is 500, so the pollution index on January 12 of Shijiazhuang and other cities was dubbed "making the meter exploding". The value of PM2.5 in Baoding even reached 632 micrograms/cubic meter, eight times higher than the current standard in China.

Officials of Hebei frequently summoned to have meetings in Beijing had never feel so stressed before.

In the steel and iron industry in China, there is a saying that "China depends on Hebei". As the largest steel and iron production province, Hebei has taken the measures to ensure the environment quality for Olympics in as early as 2008. The "joint prevention and control" measures adopted during the Olympic Games

Ecological Civilization of Contemporary China

Furnace dismantling site of Jigang Steel & Iron Co., Ltd in Zhangjiakou City, Hebei Province.

required the high-polluting companies in Beijing-Tianjin-Hebei to stop or reduce production temporarily, and demanded that heavy steel polluters within 200 kilometers away from Beijing must shut down as well. Forced by the demand in the special period of time, just in Hebei, there were 61 heavy polluters closed down before the 2008 Beijing Olympic Games, contributing to the 175 days with Grade II air quality, or better, in the first half of 2008.

Seen from the successful experience of pollution control activities during the Olympic Games to the measures the haze-torn government and enterprises were forced to take, the "joint prevention and control" and the cooperative engagement which have got back the blue sky and white clouds in Beijing-Tianjin-Hebei made a great contribution.

Presently, Beijing-Tianjin-Hebei and the surrounding regions have speeded up weeding out the outdated capacity required by the central government during the "12th Five-Year Plan" period. Although the basic principles were already given in the "12th Five-Year Plan", to implement the measures specified in the plan needs many conditions, methods and much time.

Capacity compression will certainly bring significant change to the enterprises and employees in this industrial chain of Beijing-Tianjin-Hebei, and completely change the economic pattern of Hebei. According to the public information, in the next few years, the number of iron and steel enterprises in Hebei province will reduce to about 60% of the current number year on year.

In 2013, an official agreement was entered by and between the Environmental Protection Department of Hebei Province and Beijing Municipal Environmental Protection Bureau, according to the agreement, they will establish the negotiation, report, pre-warning and linkage mechanism, promote the prevention and control cooperation for atmospheric pollution in the region of Beijing-Tianjin-Hebei, implement the *detailed rules* of national prevention and control of atmospheric pollution, and carry out trans-regional comprehensive management for air environmental pollution. Moreover, as a "hard constraint" on the steel industry in the region of Beijing-Tianjin-Hebei in the next few years, environmental protection will be connected with the rapidly tightening bank loan and become as a good tool for capacity compression.

Furthermore, the government of Hebei province has issued new policies, encouraging the iron and steel enterprises to participate in industries such as harbor, railway and sea transportation and develop strategic emerging industries, so as to transfer the steel production capacity.

It is reported that more than 5000 employees of the Tangshan Iron and Steel Group Co. of HBIS have left the plant site and engaged in non-steel industry on annual average. In Zunhua city, Hebei province, there are more than 30 of steel enterprises investing in development of emerging industries in recent two years.

The statistical data show that the crude steel production of China in 2012 was more than 700 million tons, including 329 million tons produced from Liaoning province, Hebei province, Shanxi province and Shandong province in adjacent to Beijing, which accounted for about 46% of national total production. Among all productions, steel production of Hebei province was 180 million tons, accounting for one fourth of national total production of crude steel production.

In the rank of 74 cities air quality index in the third quarter of 2013, newly issued by environment protecting department, 8 cities from the region of Beijing-Tianjin-Hebei rank the top 10 cities with the worst air quality.

Ecological Civilization of Contemporary China

On September 17, 2013, the *Detailed Rules for the Implementation of Action Plan for Prevention and Control of Atmospheric Pollution in Beijing-Tianjin-Hebei and Surrounding Area* was jointly issued by the Ministry of Environmental Protection, National Development and Reform Commission and National Energy Administration as well as other departments, and the specific capacity compression standard was made for "major" provinces engaging in steel industry such as Shandong, Liaoning, Shanxi and Hebei. The *Detailed Rules* requires that in order to support the new policy of pollution treatment, Hebei province shall compress and obsolete over 60 million tons of steel production capacity by the end of 2017, which is equivalent to reduce one third of total production.

The *Detailed Rules* also requires that the related departments in Beijing-Tianjin-Hebei and surrounding area shall not approve new capacity projects of serious overcapacity industry such as steel, cement, electrolytic aluminum, plate glass and ship.

Cement dust flying everywhere in the outdoor concrete stirring site in Dingxing County, Baoding.

As of 7:00AM on March 17, 2014, according to the latest data issued by the Ministry of Environmental Protection, the region of Beijing-Tianjin-Hebei is suffered from thick haze once more. Beijing is in middle level pollution with AQI (air quality index) of 175, Tianjin is in heavy pollution with AQI of 207, and Xingtai city, Shijiazhuang city, Zhangjiakou city, Baoding city, Hengshui city, Tangshan city and Handan city of Hebei province are in heavy pollution.

On October 14, 2013, the Ministry of Finance issued information that the central finance had arranged RMB 5 billion of funds currently, which will be completely used in the air pollution treatment work for Beijing-Tianjin-Hebei and surrounding area. The area includes Beijing, Tianjin, Hebei province, Shandong province and Inner Mongolia, and the focus is put on Hebei province with heavy treatment tasks. The Ministry of Finance states that this fund will be allocated

In 2012, the Beijing Botanical Garden, like a natural pallet, added colorful elements to the spring in Beijing.

in the way of "reward replace subsidy" and depend on the expected pollutants reductions, pollution treatment investment and PM2.5 concentration reduction proportion of the abovementioned areas.

On October 15, 2013, the State Council issued the *Guiding Opinion on Policy of Solving Overcapacity*, stating that it will effectively promote and solve the overcapacity contradictions in industries such as steel, cement and electrolytic aluminum. Moreover, as a major industry of overcapacity, steel industry will be the primary objective of management, it is planned to compress 80 million tons of total production capacity within the next 5 years.

The media have disclosed that after "receiving the order" of reducing 60 million tons of capacity, Hebei province has already resolved this index, with Tangshan, Handan and Shijiazhuang (major cities engaging in steel industry) undertaking the task of reducing 40 million tons, 12.04 million tons and 4.82 million tons respectively, and the rest will be allocated to other counties and cities.

In March 2014, the science and technological departments at provincial and municipal level in Beijing-Tianjin-Hebei have publicly collect maturely technological achievements in environmental protection such as air pollution prevention and control and water pollution control, and started to prepare the *List of the Recommended Technological Products for Atmospheric Pollution Prevention and Control in Beijing-Tianjin-Hebei*. It is reported that by focusing on key issues influencing ecology, such as air pollution, water treatment and solid wastes treatment, the related departments at provincial and municipal level of Beijing-Tianjin-Hebei will establish negotiation mechanism and regular meeting system, and they will hold seminar regularly, carrying out technological cooperation and sharing information in environmental protection, so as to provide technological support for the sustainable development of Beijing-Tianjin-Hebei.

It is an ideal situation of balancing economic development with environmental protection, which is not unlikely to realize in practice. The total population of Beijing-Tianjin-Hebei is 106 million, accounting for about 1.7% of the world total population (6.3 billion). While producing 15% of the world pig iron, they also become as the largest smokestack emitting pollutants.

All steel enterprises have to face the reality and challenge in terms of regional joint prevention and control, industrial transfer, focusing on circular economy and broadening and extending the non-steel production service industry based on steel industry.

Green Land

There is not only blue sky but also green land in ecological China.

Dreams will bloom in our mind when we enjoy free breathing in the world. It is conceivable what a terrible scene is, if there is no green on the earth as there are no waves in rivers, no clouds in the sky, and no bird singing in the forest.

Green, the most attractive one among colors, is the base color of the nature, a kind of artistic conception, and a kind of quality. So, green is not only the color of life, but also the color of dream. Therefore, beautiful China Dream shall be painted with the color of dream and burst into colorful flowers after all little efforts.

Many people must have seen green bristlegrass silently growing on the roadside of some village. Such grass bows its head down as shyly as little earhead and waves slightly in the wind. Its seeds are the food for many birds to satisfy their hunger. Especially in iced winter, birds usually rely on seeds of green bristlegasses to pass the winter. Green bristlegasses are natural

Ecological Civilization of Contemporary China

Pear Flower Village, having ten-thousand-*mu* pear garden, in Pangge Town, on the eastern bank of Yongding River in the southwestern suburb of Beijing.

earheads that the nature gives to living creatures.

Many years later, when finally finding the green bristlegrass in a modern city with concrete forests as that in the village of childhood, those who had intimate contact with the nature in the childhood will cannot help to squat and tell his long-lost miss to her.

Green bristlegrass reminds us of the country life, showing a desire for blue sky, blue water and the oneness of man and nature. Only if there is green land, can we have a true childhood and enjoy the real beauty.

In the past, people anticipated to dig out "gold" from the soil; then, high mountains were changed to wasteyard after the grand slam of cannon for mining; luxuriant and green forests were changed to wasteland by the electric saw for lumbering; the wide prairie, in which sheep and cattle were easily seen when the wind was lowering grass in green, was changed to desert…

Green Land

Goats are everywhere on the hills around the Cuandixia Village, Zhaitang Town, Mentougou District, western suburb of Beijing.

Now, all these are changing quietly. More and more people find that we can only dig out less and less "gold" if we continue digging "gold", but we can obtain more and more "gold" if we make "gold" grow from soil naturally; finally, green land are changed to gold land. And this is the really beautiful and gold China Dream.

Great Changes in Coal City: "Coal Sea" Changing Into Sea of Trees

Fragmented mountains and rivers are changed into ecological gardens; deserted and collapsed earth surface is changed into the re-plough land; cracked and dangerous shanty towns are moved to new buildings; trees and corps are planted in the earth; underground water level is improved and gush out clean water...

Could you image that "there were all sands and stones on the ground, and no trees and birds along the side of deserted land" before?

When it comes to Xishan, we often thought of Taiyuan Xishan Mining Area with a capacity of tens of millions of tons of coals every year in the past; but now, we will think of the Taiyuan Xishan 10,000-*mu* Ecological Park more.

Worries are buried deeply in the earth crust like coal, but disappear with so-called economic benefit from the moment of starting mining.

In 1980s, there was still a mined-out area filled with thick brush, garbage, fugitive dust and waste water everywhere, the withered and rotten ecological plants and the cutoff brooks among the mountains, which all have been primary pollution sources that had impacts on air mass of Taiyuan City for years.

In recent years, the People's Government of Shanxi Province took a series of effective measures. First, from 2006, the government invested RMB 6.8 billion and spent three years in governing key state-owned coalmine subsidence areas, relocating, maintaining and reinforcing damaged civil houses, schools and hospitals, making 180,000 households and more than 600,000 affected residents move into new buildings. Later, the government conducted coal resources

Flowers in full blossom in spring in the ten-thousand-*mu* ecological garden on Xishan Mountain, Shanxi Province.

reassignment and closed thousands of small coalmines and coal pits.

The magnificent changes of Xishan started under this background. In 2008, Taiyuan City removed 71 enterprises related to 6 heavily polluting industries, and closed 74 small coal mines with a capacity of less than 90,000 MMTPA, building "one mountain and seven rivers" greening system with Xishan Mountain Chain as the background and the seven rivers along the mountains as the vestibule. The government also organized to cover soil and develop land, build roads, draw water uphill, conduct mountain grazing prohibition, control and deal with pollution…wherein, people cleared away more than 20 coal yards, filled garbage of more than 100,000 cubic meters, built terraced fields and fish-scale pits, leveled up swales, and planted landscape trees and economic forests.

Xishan seems to be a rebuilt ecological pearl, making the re-greening hills and rivers attract birds to return. These attractive, fresh, brisk and poetic green

Ecological Civilization of Contemporary China

Lilacs in the ten-thousand-*mu* ecological garden on Xishan Mountain.

scenes make great contributions to citizen's happy life. When standing at the highest place of the Ecological Park and looking around, we can see pavilions, watersides, green waves in the river, and new roads up and down in the forests. What a wonderful garden scenery in mountain region!

When intoxicated with the fresh, brisk and poetic scenery of blue sky, low clouds, light wind and green forests, people easily forget that the place underfoot was a far and wide "coal sea".

Links

- With the development of large-scale of afforestation, the air quality of Taiyuan region has been improved obviously; the high wind, sand and dust weather have been decreased obviously; coal ash floating in the air has been effectively controlled, which all build a Green Great Wall of wind prevention, sand fixation and water conservation for Shanxi region when improving people's living environment.

- In 2008, Shanxi developed a five-year *Plan of Shanxi Province on Mining Geo-Environmental Protection and Management*, and launched Measures for Deposit of Mining Geological Disaster Prevention and Control. Li Jiangong, Head of the Department of

Land Resource of Shanxi Province, said, "We will accelerate the geo-environmental restoration and control of 12 key rehabilitation regions including Datong, and arrange 45 projects of mining geological disaster prevention and control with proposed investment of RMB 4.868 billion from 2011 to 2015."

At the end of 2012, RMB 220 million was invested in the Project of Greening Taiyuan Xishan 1000-*mu* Ecological Park; the afforestation of nearly 10,000 *mu* was completed; and more than 1.1 million trees in 100 varieties were planted in the Park.

It seems to be a simple ecological logic to make environment promote social harmony, use landscapes to create wealth, make ecological environment recover naturally, change "coal sea" to "forest sea" and make "coal sea" full of forests again. But, there is still a difficult cognitive process behind this, as well as a wonderful poem of harmonious coexistence of man and nature, ecology and economy.

Hero in the Desert: "Shihuichui" Fighting Against Langwosha for Three Times

"Shihuichui" (the nickname of a man) with baiyangdu kerchief around his head poured hard liquor into a soil porcelain bowl full, used his rough hand to held the bowl high, and howled like a tough man, "We will always fight against sandstorm. We do not live in vain if we complete this thing successfully!", and then touched his bowl with others' bowls and drunk off the liquor.

At that chilly cold dawn of 30 years ago, Shi Guangyin, a peasant from Dingbian County of Shaanxi and with a nickname of "Shuihuichui" did not predicted that he would be widely known for his fight against Langwosha for three times, and even be honored as the first "Hero of Desertification Control" in the history of New China by the State Forestry of Administration in 2002.

"Shihuichui" means "fool" in local slang of Shaanxi.

Why does a "Hero of Desertification Control" have such a nickname?

In Dingbian County of Shaanxi, located to the south of Mu Us Desert, the area of desert accounts for a quarter of the total area, and more than half of population are affected by sandstorm all the year around. At that time, there was a popular song that "due to the frequent sandstorm, we are usually rewarded nothing by our hard work throughout a year". From this song, it is not difficult to image what a bad environment it was at that time.

Shi Guangyin has had deep feelings about the sand damage since his childhood. When he was eight years old, he took cattle out to graze on one day, and then was blown to Inner Mongolia by sandstorm. Three days later, his farther found him 30 li (1 li = 0.5 km) away from his home. At that time, Shi Guangyin

Green Land

Desert Landscape in Anbian Town, Dingbian County, Shaanxi Province.

made up his mind that he would make all efforts to make next generations get rid of sand harm which had influenced people for generations.

In 1984, to accelerate the governance speed of barren desert and wastelands, relevant authority issued the policy of "allowing individual to contract barren desert and own the forest planted by himself". State-owned Changmaotan Forestry Station is a desertification control unit funded by the state. There were special tasks of desertification control every year before 1984, but this barren desert named as "Langwosha" had not been controlled due to highly difficult governance and ineffective management and protection, etc.

Shi Guangyin thought this was an opportunity for him, so he sold mule and sheep, and cut off all means of retreat to fight against "Langwosha".

At that time, people in his village all thought he was completely out of his mind. It's impossible to grow any seedling in boundless and indistinct sea of sands, and he was totally a "fool" who ran headlong and never come back until there was a dead end.

"Shihuichui" started off, eating steamed buns and drinking unboiled water to satisfy his hunger. "I tell you that I stayed at Langwosha day and night at that time. I insisted on not going home. In fact, I knew it's difficult to control the desertification there and it's required for a lot of manpower. At that time, the desertification of this place could be controlled, so I persuaded many people to do with me." Talking about the past, Shi Guangyin was full of emotional thoughts, "We were working hard like mule and eating food that were as awful as the food of pig and dog. At that time, a person would get blisters on his face and body and slough layer after layer under the blazing sun for a spring (forty days) there."

Even though they finally planted trees on "Langwosha", such trees were blew away by dozen of strong wind.

Such continuous "meeting a dead end" caused the failure in the first year; but he kept on trying in the second year, though he had failed for many times. "Langwosha" was as forbidding as its name, but "Shihuichui" did not give up.

Shi Guangyin cultivates pinus sylvestris saplings by himself due to the low survival rate of saplings got from other places.

Three year later, the trees stood steadily and survived! He finally obtained a return after fighting against "Langwosha" for three times.

"Shihuichui" planted trees all over the 58000-*mu* sandy land of "Langwosha". A few years later, this place became a real forest farm filled with trees and woods.

Originally, it's required to persist and march forward courageously on the way of desertification control like a fool.

Tens of years later, such seedlings in that year have grown to towering trees and the desert has been changing into an oasis little by little under governance; tens of years later, more than 200,000-*mu* forest has been planted on the barren deserts and reached a economic value of more than RMB 30 million under the governance of "Shihuichui".

The hero and big trees stand side by side on the long way of projects in the northeast, north and northwest of China (collectively referred to "three-north").

Links

- In November 1978, Chinese government made a significant strategic decision to establish a three-north shelterbelts system in the northeast, north and northwest of China where there were hazards of sand storms and severe water and soil loss. The project involved 13 provinces and covered 42.2% of national territorial area. And it pioneered in the large-scale ecological construction in China.

- In 1986, the phase II project of construction of three-north shelterbelts system was launched. In 1989, Deng Xiaoping inscribed four words: "绿色长城 (Green Great Wall)" for the three-north program. In 1996, the phase III project of construction of three-north shelterbelts system was launched. In 2001, the phase IV project of construction of three-north shelterbelts system was launched. In 2012, the State Council held a working conference on three-north shelterbelt program in Shuozhou of Shanxi; and the phase V project of construction of three-north shelterbelts system was launched.

- In 2009, the General Office of the State Council of the People's Republic of China issued *Opinions on Further Promoting Construction of Three-North Shelterbelts System*, which clearly stated that to accelerate the construction of three-north program was of great significance, and was an important measure for implementing scientific outlook on development and building ecological civilization and a strategic choice of improving eco-environment in the northeast, north and northwest of China and expanding the survival and development space of the Chinese nation.

In 2013, after 35-year construction of the three-north shelterbelts, the forest stock volume in the engineering area increased from 720 million cubic meters in 1977 to 1.39 billion cubic meters, a net increase of 670 million cubic meters.

Aspen protection forest for pasture and fields in Old Barag Banner, Hulenbeir City, Inner Mongolia.

Green Land

Look down the Shaanxi elm forest at the south edge of the Mu Us Desert in June 2011, seeing the initial effect of tens-year sand control efforts.

Various established economic forests have been 4 million hectares, while the High-quality Apple Base with Loess Plateau as the center, the Red Dates Base along the Yellow River, Xinjiang Bergamot Pear Base, Ningxia Medlar Base, Hebei Castanea Mollissima Base and other industrial belts were established. The annual output of dry and fresh fruit in the northeast, north and northwest of China is 36 million tons, accounting for one third of nationwide total output. Wherein, the annual output of apples is 16 million tons, accounting for 60% of nationwide total output.

The painstaking effort of "Shihuichui" is full of sorrow and joy of a generation devoting themselves to desertification control. It is the dauntless and persistent efforts of these "Shihuichui" that brought the barren desert to green life. 35-year efforts and decades of hardworking built today's Great Wall. It can be said that during the 35 years of construction of three-north program, the ecology has recovered and been protected, and humans have recognized the nature again.

Grassland Restoration Experiment: The "Wise Method" of Jiang Gaoming

Who is Jiang Gaoming? Perhaps, many people do not know him.

He is widely known in Chinese ecology community, and works as a research fellow in Institute of Botany, the Chinese Academy of Sciences.

Institute of Botany, the Chinese Academy of Sciences is located on the opposite of Beijing Botanical Garden; the environment there is comfortable and quiet, and very suitable for conducting academic work. Jiang Gaoming is a person who cannot stay at home and do nothing. Instead, he likes doing experiments in such a graceful laboratory and prefers to survey all over the country.

Jiang Gaoming said that there are less and less academicians and scholars who practice themselves like Yuan Longping. After all, it's easier to do experiments in a laboratory and publish articles. However, peasants will trust and follow a scientist only after he can work out specific sample plate and prove with practices.

Before and after 2000, there was a tide of afforestation on the sandy land of prairie, for which the cattle in Bayinhushu did not have grass to eat in that winter. To keep cattle alive, herdsmen had to buy grasses in places beyond 40 km, and each household spent about RMB 20,000. Having known such current situation, Jiang Gaoming started to introduce projects. At that time, with an attitude of "trying as a last resort to save a hopeless situation", herdsmen gave severely degraded saeter of 40,000 *mu* to Jiang Gaoming to do grassland restoration experiment.

Green Land

Jiang Gaoming in the top grass in grassland after the ecological restoration.

It was this experiment that brought earth-shaking changes to Bayinhushu, a pastoral village which has 70 households of herdsmen and 310 persons and covers an area of 121,000 *mu*.

Only one year later after the start of the experiment, the turf height reached 1.43m there; the yield of grass in each *mu* is 5300 *jin* (1 *jin* = 500g) in fresh weight; the seedlings of elms naturally growing reaches 321 per square meter around the elms on sandy land. In 2004, each household of herdsman could get hay of 70,000 *jin* by distribution and did not need to buy 20,000 *jin* hay like before any more, ending the history that herdsmen had to walk 40km way to buy hay. Then, and their income began to increase. All these made herdsmen wild with joy.

In the following years, the severely degraded sandy grasslands of 40,000 *mu* in Bayinhushu have been restored to its conditions in 1960s. As the vegetation recovered, wild animals returned.

After the success of the experiment, surrounding areas followed suit and the sand dunes are now covered with a healthy layer of vegetation.

What unique skills were used by Jiang Gaoming? It seemed no unique skills.

Compared with traditional means of ecological restoration such as greening and afforestation, aerial seeding, digging a well and establishing fence, the method used by Jiang Gaoming seems "to have no technical content".

His "wise method" responds to one of his important opinions of *"Nature Can Heal Itself"*.

It takes more than 2000 years to form 1cm thick soil in natural world. And the ecological system also has various functions for adapting to environment and self-healing like biological organism. In many places, the grassland does not degrade really and completely, and there are still various propagules (seed, spore, fruit, sprouting root, sprouting seedling, etc.) distributed naturally and the genes. Where its peel and ribs exist well, such propagule can revive and grow only after it rains once.

Grassland ecology revives naturally without external intervention. It is efficient and effective to deal with the contradiction between man and nature more by purposive, regular and peaceful methods.

At the end of *Nature Can Heal Itself*, written by Jiang Gaoming, there is a paragraph: "An old saying goes 'Civilian ministers ought to give good advice in defiance of death, and marshals ought to fight on the battlefield to the very end of their lives'. No one appoints me as 'remonstrator', but I would like to apply the knowledge I have learnt to the specific practice of national eco-environmental protection, and work as an 'advisory' scholar. No matter whether there are listeners, I always enjoy the joy and bitter therein. I'll search with my will unbending."

Jiang Gaoming may believe in the natural world most among well-known scientists.

Green Land

Vitality reappears in the Bashang Grassland in Zhangjiakou City, Hebei Province, after the ecological restoration.

Today, the ecology is damaged severely. Let nature heal itself without human intervention. Perhaps, the more important, practical reference significance exits in the "wise method" of Jiang Gaoming.

Links

- In 2008, Jiang Gaoming published a book entitled Nature Can Heal Itself, which takes the five-year ecology restoration experiments in semi-arid region as the clues, records his scientific research and desertification control practices in Bayinhushu for years, and states the idea that ecosystem recovers by natural force before its degradation reaches the threshold value, which caused huge controversy. In 2010, the case of Bayinhushu was included in Geography and Environment (Version 6), a college textbook of environment type in the U.S.

- After ten years from the start of ecology restoration experiment,

Horses are running freely, grass growing vibrantly.

the recovery of degraded 40,000 *mu* sandy grassland of Gacha in Bayinhushu has also driven the recovery of 120,000 *mu* grassland all over the Gacha at the same time; and the biological diversity has been protected effectively.

- In September 2012, the State Council approved RMB 87.9 billion for phase II of Beijing-Tian Sandstorm Source Project, and most of the expenditure was used for natural recovery mode, with 85% of area to recover naturally.

- A research fellow dares to give up empty talk and conduct a grassland restoration experiment that lets nature heal itself. This quality of taking up social responsibility is commendable. Ten years ago, other people in the circle sniffed at his views; ten year later, the mainstream has accepted his views.

Epitaph in a Wet Land:
The Loss and Save of "Eden"

It's frozen with cold wind in winter of 2004.

Tragedies were befalling Rongcheng Swan Lake of Shandong Province, the largest habitat of swans for wintering in Asia: for so-called requirements of economic construction, this precious natural wet land was slowing disappearing during the four-year artificial mud pumping by modern machinery.

Native people introduced that there was a natural wet land with half sand and half mud by the Swan Lake before; abundant alga and mollusks grew there; in the north, the fresh water stored by the Pinus Thunbergii Forest flows along the soft sand beach and then to the mud flat, where swans coming for wintering sought food and drank water.

But, in winter of this year, the original soft beach disappeared on the mud flat. Instead, there was thick and black rubber-like silt block covering on the beach. There were deep ditches on the surface, and only sparse withered reeds swaying with wind as far as we could see.

More than 200m away from the bank, dozens of white swans were entrenched on the dry ground and often laboriously seek water in the seam of sludge. These poor swans had waited in line for a few days and whined up for help continuously.

One swan was dead on the black sludge with wings down on its knees. Its beak were still into the seam of black sludge with slightly open eyes as if it was making the last effort to get water until it died.

In the following days, suffering from hunger, thirsty, disease and cold, beautiful white swans dropped their wings, which they used to fly in the blue sky

Ecological Civilization of Contemporary China

Thousands of swans are flying, playing and eating in Rongcheng, Shandong.

and white clouds, and laid dead one by one on the home where there was ever a lot of water and plants. Their postures of struggling hopelessly, fighting menially and complaining sadly and angrily made every person with conscience can not keep back his/her tears or at least sighed slightly after witnessing the tragedy.

Once upon a time, groups of white swans came at the appointed time for wintering and welcoming every beautiful winter; once upon a time, snows were dancing gently with a little cold sea breeze and glad to welcome every white swan to come home.

In the winter of 2004, a review article *Don't Make Swans' Beautiful Home Become Memory in the Dream* was widely read. In the article, the author wrote affectionately—

Indifference for life usually means the humanistic deficiency. We have advocated caring for animals and respecting life, and stressed that human beings are not the dominator of all things in the nature for years. But, if we cannot really care for the sufferings, life and death of non-human lives, the tender feelings of humanity cannot really return.

Perhaps, in some people's opinions, such emotion is extreme in some way. But, it is little known that human's emotion appears so subtle more.

In the busy human world, perhaps, it is the animal's emotional purity and loyalty that make human's feelings be not lost in the sea of desires and remain a little bit of trueness and simplicity. In this sense, it is the spring of emotion of animals that moistens human's withered heart.

Links

- Swans belong to anatidae of anseriformes, are national second class protected birds and a vulnerable species in the world. There are three kinds of swans in China, including whooper swan, little swan and mute swan. Those inhabiting in Rongcheng are mainly whooper swans. Whooper swan was called as "Gu" (鹄) in ancient China, belong to big swimming birds, and is a kind of birds that can fly highest in the world. It has a body length of about 140cm with the pure white feathers all over its body. Its neck is long and curve. The part from lore to mouth base is canary yellow; the part from upper beak to nostril is yellow; the part from the beak top to lower beak is black; and its claws are black. Whooper swan is living on aquatic plant, insect, etc.

- Swan Lake, called as Moon Lake geographically and located in Chengshan Town, Rongcheng City in the most eastern region of Jiaodong Peninsula, is a natural lagoon connecting to the sea with the area of 5 square kilometers, and attracts lots of whooper swans to fly from Siberia for wintering. Rongcheng Swan Lake is the largest habitat of swans for wintering in China as well as one of the four largest swan lakes in the world, and was approved as "Provincial Natural Reserve" by Shandong provincial government in 2000.

Swans are playing.

On December 20, 2004, International Fund for Animal Welfare (IFAW) conducted a one-week field visit, and issued the *Survey Report on Current Situation of Habitat of Rongcheng Swan Lake*. It's reported that the drying up of water channels and water pollution of Swan Lake were the root causes of the death of swans. The investigator of IFAW said, "We are shocked to see the tragic scene that homeless swans are seeking and fighting for food in the garbage…"

On March 8, 2005, the People's Government of Rongcheng organized relevant authorities to conduct field survey for whooper swan protection area. For problems found in field investigation, the People's Government of Rongcheng worked out five measures to strengthen the construction and dredging management of Swan Lake basin.

After the winter of 2007, due to the continuous improvement of ecological environment, the Swan Lake welcomed more than 2400 whooper swans from Siberia for wintering, while only around 110 whooper swans came during the same period of normal years. The growth rate reached 20 times.

Swans are flying.

In February 2012, under the influence of continuous snowing and cooling day after day, more than half of surface of Swan Lake had frozen, and it's inconvenient for over 1000 whooper swans to drink water and seek food. For this reason, the Forestry Department arranged corn kernels and cabbages of nearly 500 *jin* to feed them every day, ensuring that whooper swans could pass the winter safely.

In front of suffering living creatures, our menial hearts should all be the same as her. Otherwise, it's difficult for us to achieve the dignity as human beings, and even we will change to ones inferior to animals that we look down upon. It's known that animals are always considerate.

Ecological Civilization of Contemporary China

Income of RMB 11 Billion from Furs: Behind the Fur Capital

"Creak", the cage is opened and a big hand snatched at its tail into the cage.

It holds its head, shows the teeth, struggles at full split, and screams "Ha"; then the big hand loosens its tail.

It returns to the cage with head down and uneasy and aggradevole expression its eyes.

It is a raccoon dog at the age of seven months.

There are raccoons in other hundreds of same cages on open spaces of the farm. It has lived in despair since its birth. Continuously screaming like crying is

A raccoon dog trapped in a cage

the only way for it to vent. Because the destiny waiting for it is the slaughter at a certain time.

According to the habits, raccoons hide by day and come out by night. They sleep in the day, go out to find food in the evening or at night, and hibernate in cold winter. Because humans catch and kill them excessively, wild raccoons are very rare. Instead, there are many raccoons bred by humans.

Breeding raccoons has become industries in several cities and counties in Hebei, one of which is Suning County subordinated to Shangcun Market. Take Suning for example. There are 152 scaled fur animal farms, 65 specialized villages, 10,000 households and 470,000 rare animals on hand such as raccoon, fox and marten all over the County.

There was a news report *Investigation on Fur Market of Suning, Hebei: Ripping the Fur from Living Animal* released on the Internet. The report recorded the scenes that the furs of living raccoons and other animals were ripped from

Thousands pieces of rabbit, raccoon dog and fox furs hung before a fur and leather trading market in Shangcun Village, Suning County, Hebei Province

their bodies in fur markets of Hebei. The little life was screaming miserably during the whole extremely cruel process of ripping the fur—

There are locally forty or fifty such personnel specialized in ripping fur mainly from villages near the market. Each market-day coming, they will generally constitute a team of four or five persons and appear on the market on time.

According to the data from Propaganda Department of Suning County Party Committee, the fur industry has become a pillar industry for Suning to enrich the County and its people. By fur business, the financial output of Suning County increased from RMB 80 million in 2002 to RMB 11 billion in 2012, achieving a new highest record historically.

Local labor services are hardly exported in Suning County. According to open data, in 2005, there were 50,000 persons specialized in work about fur industry among the 330,000 population of the whole county.

There is a local slogan: "Chinese furs are best in Suning, while best furs of Suning mainly come from Shangcun Village."

Links

- Raccoon is a kind of relatively precious furbearer whose fur is long, light and tough with soft hair. The velour leather without guard hairs is the superior raw material for making fur coat. The guard hairs have a good elasticity, and are applicable to making paintbrush.

- Suning County is subordinated to Cangzhou City of Hebei. The fur industry in Suning forms development pattern with main industries of furbearer breeding, market gathering and distribution, dying original fur with saltpeter, fur coat processing, garment and workpieces, and earning foreign exchange through exports. Suning was identified as the "Fur Base of China" by China Leather Industry Association

Green Land

Raccoon dogs are valuable furbearers.

and as the "Provincial Fur Garment Export Base" by the People's Government of Hebei Province.

On April 4, 2005, relevant person from China Leather Industry Association indicated that it's required a process when pursuing the humanistic slaughter due to the limit of developmental level and relevant conditions. For this reason, the Association sent circular to relevant governments, local associations, breeding enterprises, fur manufacturing enterprises, etc. to propose: if there are improper methods of breeding, slaughtering and transportation during the process of breeding, please strengthen management and conduct self-examination for non-standard behaviors.

On December 16, 2007, the Launch Ceremony of Chinese Humanistic Slaughtering Plan was held in Henan. Since 2008, the humanistic slaughtering trainings have been started throughout the country.

It is a consensus for a civilized society to forbid killing animals with

Ecological Civilization of Contemporary China

Fur and leather trading market in Shangcun Village, Suning County, Hebei Province. Almost every household is engaged in fur and leather industry.

maltreatment. The attitude of citizens in a country toward animals is an important symbol for measuring the level of social civilization. The attitude toward animals directly reflects people's basic attitude toward life. At present, it is not optimistic for phenomena that animals are treated cruelly and killed with maltreatment. Related public figures call for the state to issue regulations on benefits of animals.

Clear Water

As the saying goes "water is the source of life", but in fact, people usually neglect this fountain of life.

People like to picture the life of flying in the blue sky, swimming in the blue water, running on the mother earth, what a poetic life it is! However, it is just a dream in most cases. In reality, many rivers have been polluted and cut down, and we are unable to see them flow freely with beautiful waves. On both sides of the river bank, the scent of rice blossoms across a thousand miles is replaced by factories and property development.

Faced by this severe fact, the action of "regaining clean water" is not only the direction taken by the Chinese governments, but also the goal for the Chinese people. We can make our dream come true and feel happiness only on the basis of "clean water". Moreover, "clean water" is not only

Ecological Civilization of Contemporary China

about "purifying" water, it is more about "activating" water. In rural area, activating water means that we should let the rivers flow without constraint and roar the sound of nature. In urban area, activating water means that we should be intimacy with water, making it as a real valuable platform carrying the city happiness instead of overexploitation of underground water, resulting in land subsidence and sinking our future.

Since 2000, a campaign for water has been carried out quietly in civil society. At the end of 2003, under the strong pressure from NGO, mass media and social opinion, the Yangliuhu dam project was canceled by the government, which was supposed to be built near the world natural and cultural heritage—Dujiang Weir. Focusing on the Nujiang River hydroelectric development, NGO and other social powers questioned and criticized the out-of-order hydroelectric development in the southwest region. By paying attention to the ecological effects of dam and the

Ahai Power Station on Jinshajiang River waiting for the environment impact assessment of 2008.

Fish swimming in the water in the wet land of Olympic Forest Park in Beijing.

resettlement issues, NGO has shown its more profound cognition of social attributes of ecological and environmental issues. With the continual debates over hydroelectric development of the west region, it not only promotes the government to make scientific decision on hydroelectric development and dam construction, but also provides multiple information and various comments, so as to form public participation and balance mechanism.

The "11th Five-Year Plan" proposal was issued by the central government in October 2005, and the wording of hydroelectric development was modified from "active development of hydroelectric power" to "orderly development of hydroelectric power based on protection of ecology", the in-depth concept-changing is evident. In April of the same year, the public hearings for seepage-proofing project of Old Summer Palace lakebed was hosted by the State Environmental Protection Administration at that time, which was a ground-breaking case in the environmental protection history of China. There were nearly 10 representatives from the civil environmental protection organization of NGO out of 73 representatives participating in the hearing. This event is a successful case in the cooperation between the environmental protection organization of NGO and the governmental environmental protection departments.

Until 2013, the concept of "ecological red line" was proposed at the Third Plenary Session of the 18th Central Committee of the CPC. This red line will not only draw on the land, but also apply to sky and rivers. With this "ecological red line", we believe that under the extensive participation of the governments, NGO and the public, the vision of "facing the sea with spring blossoms" will no longer be a dream, we can certainly make it as a life scene one day.

Cancer Village: Death Village under the Shadow of Pollution

It is not exaggerated to say that people daring to drink tap water of Jinling Town are not afraid of death.

A river is on the west side of Jinling Town, Zibo City, Shandong province, on February 21, 2013, its color has changed into weird verdigris with unutterable stink, which was a crystal clear river in the memory of villagers.

There is a strange story in the village: an old man used to live near the plant site. One day he went to toilet in the middle of the night, but didn't come back for a long time. His family found him falling in a faint for the strong scent

Waste water without being treated is discharged into the river outside the city.

of chemical waste gas. After being sent to the hospital, the old man was dead already.

In fact, Jinling Town has suffered from water pollution for more than 20 years. The water quality has deteriorated significantly since the establishment of ethylene industrial park on the south side of the village in around 1988. As of the establishment of Qilu industrial park in 2006, the small town has been surrounded by chemical plants in all sizes, and the pungent odor from the stream is so strong that the villagers dare not open window.

Many plants have discharged the waste water directly into the stream, polluting it into blue, red and black. "It makes me feel creepy from a distance" one villager said.

More and more people have cancer, such as gastric cancer and lung cancer. "It is like a curse". The patients all go through the same process — feel uncomfortable at first, and then have body check in hospital and identify cancer by doctor (most of them are terminal cancer) and unable to carry out surgery. The condition deteriorates quickly and the patient usually passes away within one year.

The daughters-in-law are afraid to go the mothers-in-law or mothers' homes of the town. Even coming back home, they usually stay no more than one day, and do not use water for showering or eating in particular. In the case of selling vegetables in neighboring town, people run away when knowing they grown in Jinling Town. When knowing the patient is coming from Zibo, doctor of tumor hospital will ask him/her the question "are you from Jinling?" without hesitation...

The name of cancer village has spread for miles, and the shadow of death hovering around the village.

Almost every family in the town has cancer patient, and Jinling Town has become lifelessly. The similar repeated symptom and death list are growing

Clear Water

Urban border, waste water dumping, and geese having no way home.

every day. "The cemetery is full and more and more new graves mounds appear on land", one elderly man said.

The title of "cancer village" also hurts villagers' feelings.

Until now, the water pollution situation in Jinling Town is still shocking and has not been solved.

Although the neighborhood committee and village committee have already supplied bottled mineral water to the villagers in a centralized way, but the villagers are all confused about the future of the town and their children.

In the concentrated place for mineral water, the villagers come here and queue up on time, and it has become as a natural place for information collection.

Nonetheless, most of the information they get is about death and who is identified as terminal cancer, making them feel sad.

The official statistical data show that the financial income realized by Jinling Town in 2006 was RMB 13.88 million, increasing by 56% compared to the previous year. The data has increased rapidly in the next few years, and the financial income of Jinling Town in 2012 has amounted to RMB 109.67 million, which was 8 times to that of 6 years ago.

Obviously, Jinling town has paid a painful price for the data.

Links

- Since May 1, 2012, the *Reporter Investigation* has launched a series of investigation, interview and reporting activities of "entering into the cancer village of China". According to the statistics, the number of cancer villages disclosed by the media "is nearly more than 200", and it is only the conservative number, we still do not know the exact number of cancer villages in China.

- The *Chinese Cancer Registry Annual Report of 2012* was issued on January 9, 2013. It shows that in the recent 20 years, the cancer has showed an increase trend in three aspects: patients are younger in average age, higher in morbidity rate and death rate. There are about 3,120,000 new cases of neoplastic disease every year, which are averaged to 8550 people getting sick every day, and 6 people are identified as cancer patient in every minute nationwide.

- The Ministry of Environmental Protection issued the *12th Five-Year Plan of Prevention and Control on Environmental Risks of Chemicals* on February 21, 2013. The Plan states that more than 3000 kinds of chemical materials in China severely harm the human physical health and ecological environment, causing serious health and social issues, even "cancer villages" appeared in some regions. The Plan states that the central government will carry out comprehensive control and treatment for the chemical pollution during the "12th Five-Year Plan".

- Digital version of the *Water Environment along Huaihe River Basin and Gastrointestinal Cancer Death Atlas* was published on June 25, 2013, which is the achievement of long-term research by the professional team of the Chinese Center for Disease Control and Prevention. It has verified that the high rate of cancer is directly related to water pollution.

In Wuhu, Anhui, a girl in red dabbling in the square.

Because water is the source of life, the issue on the live bottom line of cancer village caused by water pollution is inevitable. The waste water under the ground of cancer village is still flowing, and the villagers are still struggled for living in the toxic water. The economic development is exchanged by sacrificing ecological environment, and the GDP is achieved with "blood".

RMB 8 Billion of Salvation: Truth of Dead Fish in Baiyangdian Lake "Surfaces above the Water"

The boat dips into the waves along the waterway, going through the high cattail, reed and a cluster of lotus on both side of the boat. We can hear the tweet of one or two frightened waterfowls flying away.

Blessed with the clean lake and fat fish and as the largest the freshwater lake in the northern China, Baiyangdian Lake is the beautiful "pearl of North China", and it has appeared in many movie and television plays.

However, one day in March 2006, Zhang Hongda, the local resident in the area of Baiyangdian Lake, was down in spirit. The lake, which was once a proud of Zhang Hongda, was floating with many small and big fish on the black water, and the pungent stench could not be blown away by the cool wind in the morning.

Ten years ago, Zhang Hongda and the other three men contracted this water area which was collectively owned by the village. They enclosed nets in the surrounding area and invested a lot in operating fish farm. Because 2006 was the previous year of contract, he specifically didn't fish in 2005, and prepared for the excellent harvest for the last time. According to the rate of survival of the year and fish amount per year, the harvestable fish was calculated as 50,000 kg. However, within this two days, he fished out nearly 4000 kg dead fish in a raw within the contacted water area (140 *mu*), indicating that all his work done in this year was wasted.

In the early March, the ice surface of Baiyangdian Lake just begins to melt. According to the local traditional customs, it was the ideal opportunity for

Ten-thousand-*mu* reeds in Baiyang Lake.

the fishermen fishing in the Lake. In fact, before the ice began to melt, Zhang Hongda had found out the situation of partial dead fish, but because it was very common for the Lake, he thought he could gain some fish by luck at that time.

The fishermen are living by making use of local resources. In the past, the income from fishing was sufficient to maintain the basic living standard of a family. However, the current situation was worse than that of imagination.

No live fish in the Lake, not even one, Zhang Hongda was desperate.

It made people fall into desperation after the tragedy of "dead fish event" happened in 2000. People are depressed by the pollution treatment and salvation plan, which is a 10-year plan from 2005 with total investment of RMB 8 billion.

In a series of control pollution action carried out resolutely, by the resolute measures such as cutting off water or electric supply, the governments immediately closed 142 industrial enterprises that had simple and crude pollution prevention and control facilities and were unable to secure the long-term compliance with the standards in a stable manner. Moreover, 4 officials have received punishment for ineffective control of pollution.

The agricultural monitoring station also presented the final inspection report for verifying the causes of dead fish. Zhang Hongda and the other fishermen received compensations from the Baoding municipal government by this report. However, the comfort with "RMB 1000 for 2000 *jin* fish" was far from recovering the costs of fish farm for Zhang Hongda.

Dramatically, a few years later, the waste water was still continually discharged into the Lake.

On August 12, 2012, partial water area in the Baiyangdian Lake appeared the phenomenon of dead fish in large amount again. Within three days, fish in 2000 *mu* of water area all died. Compared to Zhang Hongda losing everything, this event made major fishermen of fish farm lose millions of costs, because they had experienced dead fish events for several times and continued to invest in fishing, making them go to bankruptcy.

"The fertile land of fish and rice" in the past became as "the land of dead fish" at the moment.

At present, many fishermen in the villages have given up the business of relying on fishing, leaving their hometown and working in the city.

Links

- Baiyangdian Lake suffered from the incident of dead fish in large area in 2000, which was resulted from the water area pollution of the Lake. The Ministry of Agriculture identified the incident as "extra serious pollution accident of fishing industry". Anxin County suffered from economic lose with RMB 23.8592 million, including RMB 9.0758 million for breeding industry and RMB 14.7834 million for natural fishery resources.

Clear Water

On Match 22, In March 2006, despite the huge labor and financial efforts local government had devoted to the ecological environment treatment, there is still a large amount of dead fish floating on the Baiyang Lake.

- On December 18, 2005, the Hebei provincial government announced to the media that it had started to launch an ecological environment management plan for the Baiyangdian Lake, which was a ten-year plan with RMB 8 billion of total investment. The plan is expected to make the Baiyangdian Lake—the kidney of North China regains vitality.

- On March 21, 2006, in terms of the matters about dead fish appearing in large water area of the Baiyangdian Lake after the "dead fish event" of 2000, Baoding municipal Party committee and the municipal government publicly issued the *Circular on the Management of Lax Enforcement of Illegal Pollution Discharge by Paper-Making Enterprises in Xinshi District*. The Circular states that due to the ineffective control over pollution, one competent vice director of the Xinshi District received punishment of administrative

Ecological Civilization of Contemporary China

Fragrance of lotus in the vast Baiyang Lake can be smelled hundreds miles away.

disciplinary warning, director of environmental protection bureau of Xinshi District resigned to assume responsibility and two vice directors of the environmental protection bureau were removed.

Destruction on the environment is irreversible, and the cost for saving it is more expensive.

We need to develop economy while protecting the green hills and clear waters, which is the ecological environment protection slogan publicized by many cities. However, when facing the objective facts and making a choice between the economic development and the environment, how many people is able to resist the temptation of profits and really insist on guarding the green hills and clear waters?

Events of the Baiyangdian Lake should be like a mirror, from which people can examine their behavior and understand what really matters.

Public Benefit Database: the Wisdom of Water Pollution Map

In September 2006, as director of the "Institute of Public & Environmental Affairs" from the environmental organization NGO, Ma Jun opened the first environmental protection public benefit database in China in high profile — the website of "China's water pollution map" (www.ipe.org.cn). The original intention of the map was a database and communication platform for the professionals and organizations of environmental protection research, it has drawn attention from the social public for presenting the pollution situation in the form of map, which is novel and straight forward. As a result, the click rate of the website has increased rapidly and become as a platform for the internet users reporting water pollution in a short period.

Clicking on the homepage of the website, you will see a complete map of China's administrative divisions shown on the screen. The map collects the statistical data from departments of environmental protection, water conservancy, sea, land resources and fishery and so on, covering the information about water quality, pollution discharge, pollution sources, enterprises exceeding the discharge standards and waste water treatment plants from 31 of provincial-level administrative districts and about 300 of prefecture-level cities.

Over ten thousand of polluting enterprises are reported and exposed to the public by this platform, including more than 270 of well-known multinational companies which are famous for environmental protection.It is a very important exploration in the history of environmental protection to "hunt down" the water pollution sources and polluting enterprises on the internet.Any one who are willing to do this and concern about China's environment can participate in the action of exposure and report.

Ecological Civilization of Contemporary China

After clicking, you will see in the pollution of surface water sections in China, only 3% of water is relatively less polluted and clean, and 60% of surface water is severely polluted. Regions like Yangtze Estuary, Hangzhou Gulf and Pearl River Estuary and the offshore in particular are under very serious pollution, because 80% of sewage draining exiting to the sea from the enterprises nationwide has not reached the discharge standards. More seriously, some regional pollutions in red and black area on the map are rivers reaches under relatively sever pollution, and the black stands for the water that cannot be used anymore. It is shocking to see the frequent appearance of reaches in black.

In 2006, Ma Jun was elected as "Top 100 of The World's Most Influential People" by the *Times Magazine* along with Weng Jiabao (Premier of the State Council), Huang Guangyu (Chairman of the Board of Gome Group) and Ang Lee. Ironically, on May 18, 2010, Huang Guangyu was accused of illegal business operations crime, insider trading and unit crime of offering bribes by the Beijing Second Intermediate Court, and was sentenced 14 years of fixed-term imprisonment at the first trial, with penalty of RMB 600 million and expropriation of RMB 200 million. After it, we do not know how the world will be influenced by him, but Ma Jun and his co-workers from the environmental organization NGO are influencing the world further.

The *Times* Magazine says—

> If we put the head portrait of Ma Jun with that of the basketball star Yao Ming and the beautiful movie star Zhang Ziyi on the billboard in the streets of Beijing, most of the passers-by certainly do not know who he is. However, for those who know Ma Jun, they will say that China needs people like him, and the demand for this kind of people is far more urgent than that for a sport super star or a movie star.

Actually, since global environment is deteriorating, both China and the world need the environmentalists like Ma Jun.

Clear Water

Ma Jun, Director of "Public and Environment Research Center", a environmental protection NGO.

Before the election, the *Times* Magazine gave a high praise for the *China's Water Crisis* written by Ma Jun, and it commented that "the meaning of Ma Jun's *China's Water Crisis* to China maybe just as Rachel Carson's Silent Spring to America."

What does the Silent Spring mean to America? Rachel Carson published the Silent Spring as early as 1962. At that time, the book was regarded as frightening prediction about pesticides harming human environment, and it was strongly criticized by the interested production and economic departments, and also shocked the general public. One year later, the American president Kennedy at that time appointed the special committee to investigate and collect evidence, finally confirming the conclusion of the book that the pesticides had potential hazards. After the close of the Congress hearing, the first civil environmental organization was formed, and the US Environmental Protection Agency was established under this background. DDT(Its inventor was awarded by Nobel Prize for it) and several other highly toxic pesticides were finally completely removed out of the production and application list.

Although the *China's Water Crisis* has become the essential reference book for many overseas researchers researching on China's water issue, but it has not yet completely played its influence power on China today.

Until now, China is still facing the serious issue of water pollution, and the prediction of *China's water crisis* is coming true gradually.

Links

- Institute of Public & Environmental Affairs (IPE) is a NGO environmental protection institution in Beijing. Since establishment in May 2006, IPE has developed and operated the database of China's pollution map, promoted the publication of environment information and facilitated the improvement of environmental management mechanism.

- The year of 2012 is the 5th year after the issuance of *The Regulations of Government Information Disclosure* and *Environment Information Disclosure Methods (Trial)*, and the environmental protection departments at state and local level have added the normative documents related to the detailed matters on issuance and publicity key points of environmental information. Moreover, they have also made a breakthrough in environmental law enforcement.

- In October 2012, by virtue of the database of "China's water pollution map", Ma Jun won the Third China Soft Science Prize.

Launch of "China's water pollution map" and "hunt down pollution on internet" have effectively promoted the interaction between governments, enterprises and the public. On the one hand, it allows people to experience the current severe situation of water pollution in China in a more direct way; on the other hand, it has guided the issue of public participation and pollution treatment in depth and practicability.

Ecosystem Assessment: "Eight Doubts" for Lake-lining Project of the Old Summer Palace

The Old Summer Palace is located at western suburb of Beijing, adjacent to the Summer Palace.

On March 22, 2005, tourists from all over the world were as always sightseeing in this the royal garden (also being called "Garden of Gardens" in China) and Zhang Zhengchun, a professor from School of Life Science, Lanzhou University specialized in ecological study was one of them.

And this day was also World Water Day.

Around by people who were laughing and talking happily, Prof. Zhang was shocked by another "white" view:

Dozens of noisy excavators were busying flattening the renovated lakebed in the Old Summer Palace while workers are pressing layers upon layers plastic films onto stones.

Characterized with width of 6m and length of 50m, this impervious white barrier was formed with one layer of thicker plastic film and one layer of mantle. At first, those excavators dug out the sludge from the lakebed and then laid soil layer of 1 meter onto the ready plastic films. Workers who were taking a rest at the lake told Prof. Zhang that Lake-lining Project of the Old Summer Palace started from February 16, 2005 and for now the whole project was almost finished with all the antiseep plastic films laid.

It was said that it was to protect the lake from penetration. It was based on the fact that workers of the Old Summer Palace would have to fill the lake for

Ecological Civilization of Contemporary China

three times the last year, however, with the plastic films laid, only one fill would be okay, which would save about millions of Renminbi. It was also said that this large-scale plastic film laying costed more than RMB 120 million.

All of the details seen from Prof. Zhang's eyes made him stunned. He can tell from his common sense that this is absolute damage to the ecological environment and to the cultural relics.

It is vitally important for this open ecological system of the park to keep harmonious relationship with its surroundings. As regard to the Old Summer Palace, sustainable development and ecological balance have stood tests for hundreds of years. In the aspect of ecology, the exchanges of substance, energy and information are dispensable. Exchanges between water system in this park, in external environment and under ground formed the most important ecological flow. In this sense, water system in the Old Summer Palace is its lifeblood and if it is damaged, its lifeblood is gone.

Main work of the lakebed seepage-proofing work for Yuanmingyuan Ruins Park in Beijing.

This impermeable plastic film is very harmful to the environment, which is also known as "white pollution". However, it now wraps the lake water to create an oxygen deficiency environment where anaerobic livings release a large amount of poisonous gas, which is harmful to biocenosis, fishes, waterfowl and tourists. And it also turns this lake "dead water".

Not just keeping the fury in his chest, Prof. Zhang called the media to make it public.

On March 30, management committee of the Old Summer Palace, insisting that this project was necessary, gave their reply on the "impermeable plastic film issue": impermeable project shall not be included into the construction work category; severe water shortage situation should be lifted without doubt; project proposal had been reasoned for several times; impermeable project would not have any influences on the environment.

However, this reply stirred up a big uproar in the public and brought up eight doubts:

First of all, "the Old Summer Palace boasts to be an artistic masterpiece in the heyday of Qing dynasty", but in fact workers treat the stones and bricks dismantled from the bank in a very casual manner. So, may I ask in which way you are showing cherishing and respect for this masterpiece created by the ancients?

Second, "Lake-lining Project of the Old Summer Palace is designed to protect the environment", but environment involves four aspects, namely air, water, soil or biodiversity. With regard to the damages, I'd like ask: which aspect is the project designed to protect?

Third, "Garden of Eternal Spring is in acute water shortage situation… we have no choice but to implement this project", however, on the other hand, we can see that large-scale concrete placement dock is on progress in this place to lay groundwork for boat sailing. So, how do you explain the so-called water saving?

Zhang Zhengchun, a professor from Faculty of Life and Science of Lanzhou University, the first person who realizes the issue of Yuanmingyuan Ruins Park.

Fourth, "it will cost us more than RMB 20 million to keep water in this lake at 1.5m", but we cannot see why this lake should keep at 1.5m. So, what is the 1.5m for, for saving water or for tourist boat?

Fifth, "management committee of the Old Summer Palace has invited related organizations and experts to conduct intensive study…", but in fact all of the water-saving irrigation landscaping patterns (arbor and shrub pattern) have been changed into water deprivation landscaping patterns. So, may I ask in which way this can save water?

Sixth, "management committee of the Old Summer Palace has established a rainfall flood utilization project proposal", but as we all know, the most effective way of using rainfall is letting them penetrate in a natural manner and thus turn into underground water. So, since all of the water in the Old Summer Palace has been separated from the soil, how can the rainfall flood be used?

Seepage-proofing film at the bottom of Fuhai Lake in the Yuanmingyuan Ruins Park.

Seventh, "lake-lining project will only apply to ground floor rather than all side faces." However, in fact all joints have been blocked by concrete, so that trees at the bank have no way to absorb water from the lake. So, why knowingly violate?

Eighth, "maintain and preserve aquatic plant, creature and microbe bacteria in the lake as to find balance for the water ecosystem", but in fact all species (including aquatic plant, creature and microbe bacteria) living in the water have been completely ruined by this project. So, the doubt I have here is whether this project is trying to protect or trying to damage the ecological balance?

On April 13, State Environmental Protection Administration (SEPA) launched a public hearing on environment influences of lake-lining project of the Old Summer Palace. This hearing is unprecedented and marks a great leap no matter in the history of China's environmental protection or in the history of law.

Ecological Civilization of Contemporary China

It is undisputed that public opinion contributes a lot to this "eight doubts", where the combination of disclosure of the government and the voice of the public provides a good example of interaction between the officials and the civilians.

Links

- The Old Summer Palace (a popular name in China is Yuan Ming Yuan) is named by Kangxi, the most impressive Qing emperor. And the horizontal inscribed board bearing his writings is hung on the chamber lintel of the Old Summer Palace. Emperor Yongzheng of Qing dynasty once interpreted this name (Yuan Ming Yuan), he said that "Yuan" means one possesses all the virtues that he oversteps others and "Ming" means the outstanding performances that an emperor achieves. In one word, "Yuan Ming" is the standard measuring how distinguished an emperor and an official can be.

Beijing Yuanmingyuan Ruins Park

Peach in bloom and willow green, the spring in the Yuanmingyuan Ruins Park is so charming.

- In the beginning of 1990s, the only water remaining unopened in the west of the Old Summer Palace is the east section of the Qianhu Lake and the Houhu Lake, where has severe percolation. After entering the middle and later periods of 1990s, water shortage in Beijing worsened the water situation in the Old Summer Palace. After the year of 2000, Fuhai Lake as well as lakes in Elegant Spring Garden and Garden of Eternal Spring suffers from water shortage in most of the time even that artificial water filling is done every year.

- On September 1, 2003, the Environmental Impact Assessment Law of the People's Republic of China took effect, stating that construction projects having influence on environment without plan draft concerning environmental implication shall not be approved by related authority; projects without approval of environment protection and relevant permissions shall be shut down and banned.

- On March 31, 2005, SEPA called off lake-lining project of the Old Summer Palace with immediate effect. On May 9, 2005, SEPA delivered an ultimatum saying that management committee of the Old Summer Palace should submit environmental impact assessment (EIA) report in 40 days from now on. On July 5, 2005, SEPA released the EIA report concerning lake-lining project of the Old Summer Palace on the official website. The click rate on this website climbed to 17,000 in 10 hours, running this site to a crash for about 3 hours. On August 15, 2005, Rectification Project concerning lake-lining in Fuhai scenic region started. On July 12, 2005, SEPA delivered rectification documentation to management committee of the Old Summer Palace. On July 12, 2006, Rectification Project concerning lake-lining in the Old Summer Palace was identified as qualified by SEPA and thus put an end to the "impermeable plastic film" incident.

It is common sense that park, lake, river and wet land common sense cannot be absent from a city because park is the lung of a city while lake, river and wet land are the kidney. Harmony between human and nature as well as between economy and culture is the guarantee of a harmonious society.

Pan Yue, vice minister of Ministry of Environmental Protection, once said that "public engagement is best way explaining what is socialist democracy and in this regard, public engagement marks the best entry point." Taking a look at the "eight doubts" for Lake-lining Project of the Old Summer Palace, we will see that sound democracy system can boast to be a guardian for environmental protection.

Ten Years of Protection of Nujiang River of Wang Yongchen

"I'm driven to pursue more blue sky and more beautiful nature. I'm an environmental protection enthusiast and I want to spread this gene to more people", this is from Wang Yongchen.

It is many people's dream to live in a poetic environment. However, the increasingly severe environmental crisis in current China breaks their dream.

It is in this kind situation that Wang Yongchen insists on having a dream like this. More importantly, she put this dream into practice as an environmental protection enthusiast. She established a NGO called "Green Homeland" to pursue this dream rather than do it on her own.

As a dream-seeker for poetic living environment and a journalist in CNR, she always has the chance to do interview all around the country, which makes her one of the first batch of people who concern the environment protection during the period of high-speed economic development in China.

From 2003, hydropower station construction in Nujiang River attracted attention of "Green Homeland" and Wang Yongchen.

It was the first time for her to be that close to Nujiang River. Pure white snow and the setting sun enhanced each other's beauty and they also reflected into the First Bay of Nujiang River. Some country folks relaxed themselves with singing after one-day spent in the fair. Cheek to cheek, two women rounded their arms onto each other's neck to drink from the same bowl with their heads wagging and you can hear chants from their mouths. It was the local custom: folks who come from the fair with money earned share drink with people went

Ecological Civilization of Contemporary China

Wang Yongchen in the reservoir.

together, which means they are of one heart and one mind and it's because this, the drink is called "one heart drink".

Since then, Wang Yongchen has fallen in love with Nujiang River. She realized that once this hydropower station was launched, Nujiang River was about to rise, which means Nujiang, the World Natural Heritage, will be forever changed.

Hence, she held "Nujiang River Photographic Exhibition" by raising money on her own to make more people know Nujiang River. In this exhibition, she employed pictures to show people the beauty of Nujiang River and cultural inheritance among the 22 minorities in this area. During the National People's Congress and the Chinese Political Consultative Conference (NPC & CPPCC), she tried to make proposals to NPC member by all possible channels.

Like she said, Nujiang River not just belongs to China but belongs to all human beings. We have a lot of choices concerning electricity generating, however, we just have one Nujiang River and once it was destroyed, it would be gone.

And after all the fighting of environmental protection NGO as well as Wang Yongchen, the outcome didn't disappoint them. On February 18, 2004, Wen Jiabao, Premier of the State Council gave his comment "for issues arousing great concern from the public and bringing about different opinions on environmental protection, we shall study carefully to make scientific decision".

Wang Yongchen learnt about this from her cell phone and cried out when she was walking through the Valley of Nujiang River, exhausted. The success of protecting Nujiang River marked a milestone for China, where the voice from NGO exerted a great influence on central government's decision making and this was the first time in China.

However, things would never go as people wish. In October 2007, Nujiang Prefecture government drafted and issued *Temporary Transitional*

Ecological Civilization of Contemporary China

The first beach of Nujiang River.

Scheme for Migrants Production Arrangement of Left Bank Construction Area of Liuku Hydropower Station, saying "governments of each level shall pay attention to that the preliminary work of Liuku Hydropower Station shall not be terminated although the official approval hasn't delivered to us." In April 2009, Premier Wen Jiabao again delivered his comment on this issue "since it will arise great concern, we should make careful decision on basis of intensive opinions collection and in-depth evaluation. We should learn our lessons from hydroelectric development in Jingsha River since lots of problems left unsolved there. In light of hydropower construction, I have already made my opinion on one of reports from National Development and Reform Commission (NDRC) on October 20, 2004 for your reference."

The fact that Premier Wen Jiabao gave his comments on this for twice showed us that this issue exerted great influences on it.

For now, this seesaw battle is still going on: the construction of Nujiang

Dimaluo Dam on the branch of Nujiang River is under construction.

Ecological Civilization of Contemporary China

Landscape on both shores of Nujiang River.

Dam starts making its appearance quietly and tunnel for Songta Hydropower Station has already gone through the mountain with vehicles running in.

In 2013, for the 12th time, Wang Yongchen came to Nujiang River and she told us "Ten years, we have protected Nujiang River for ten years. For each time seeing it flow as it is created, I feel grateful."

Links

- In July 2003, Nujiang River was declared as World Natural Heritage by UNESCO.
- In August 2003, NDRC passed the proposal to build thirteen-cascade hydropower station along the middle and lower reaches of Nujiang River. According to the design proposal, this two dams and thirteen-

cascade hydropower station along the middle and lower reaches of Nujiang River located in "Three Parallel Rivers" region, which is the World Natural Heritage, have a total installed gross capacity of 21.32 million KW while the annual energy output is of 102.96 billion KWh, which is bigger than that in the Three Gorges.

- On June 18, 2006, a group of environmentalists filed a lawsuit against SEPA concerning Nujiang River issue, where all of them signed their names on the administrative proceeding documents. This was the strongest query from the civilians since Zhu Guangyao, Deputy Director of SEPA, announced big adjustment shall be made concerning Nujiang proposal on June 5, which was also the first time for the Nujiang proposal to become a lawsuit for three years.

On January 23, 2013, General Office of the State Council issued The 12th Five-Year Plan for Energy Development, where stated that the 12th Five-Year Plan would promote the development of hydropower in an active manner, in which Songta Hydropower Station at Nujiang River shall be the highlight by orderly starting Liuku, Maji, Yabiluo and Saige projects along the main stream of Nujiang with in-depth evaluation. By this, we can say that another round of hydroelectric development of Nujiang River was started.

Pursuit of poetic living environment shouldn't just be a dream of Wang Yongchen, but the environmental protection consensus of all Chinese because it is not only the core value of life, but also the inalienable right given by god.

Pollution Control for Huaihe River: The Shocking "Thirty-Six Strategies" for Illegal Pollution Discharge

Huo Daishan is definitely a notable people whenever this name is brought up in local.

And that's because his name is connected closely to Huaihe River since he has been doing things for Huaihe River protection for years. A journey of over 20 counties and cities along Huaihe River through over 4000km, more than 10,000 pictures and multiple times of large-scale environmental protection exhibition, these are all his actions for protecting Huaihe River. He revealed us the real

Huo Daishan shows the real condition of the pollution of Huaihe River.

Clear Water

Flowers Resist Pollution (shot by Huo Daishan).

situation in Huaihe River with his camera. He was in 2004 awarded as one of the ten people who value the public advantages.

In the winter of 1999, Huo Daishan found children wearing masks for class to protect them from bad smell sent forth by Yinghe River (one of the branch of Huaihe River), and the school is less than one hundred meters from it. He was so shocked and pressed the shutter immediately.

It was this picture entitled *Denial from the Flowers of This Country* that exposed some lies of pollution control in Huaihe River to the public.

On June 5, 2000, the World Environment Day, CCTV invited a mask-wearing girl from the picture to tell the public the crisis brought by the water pollution. Two days later, one of the vice provincial governors of Henan arrived at this school to make investigate and survey without pomp. This vice provincial governor made his decision on the spot that the government shall dig deep wells as soon as possible to provide villagers tap water.

In order to clarify the facts concerning the pollution in Huaihe River, Huo Daishan quitted his job to do this with his full heart. He traveled through Shiren Mountain and Shaoshi Mountain of the upper reaches of the Yinghe River, Luohe and Zhoukou of the middle reaches and Anhui and Jiangsu of the lower reaches with his camera, during which, he not only took pictures, but also collected facts about lies of pollution control along the river.

For example, some businesses directly extract water from underground to fake the scene to the relevant authorities that they had done the pollution control effectively; brought fresh fishes that weighted over one *jin* from the fair to fake the scene that water ecology had changed; or even bought fresh water from the upper stretches to fill into his section to cover the real situation. Therefore, a saying went around along both sides of the river about this: take fresh water filling as a signal for authorities' coming.

Over time, Huo Daishan found that many businesses along the river had their hidden drain outlets to avoid the inspection from the authorities. However, these hidden drain outlets cannot escape the eyes of local folks.

Huo Daishan even made a summary of the shocking tricks employed by these evil businesses and called them "Thirty-Six Strategies" of pollution discharge.

Seeking Cover Strategy: businesses built hidden drain outlets under the cover of building sewage treatment works so that the sewage can discharge directly without processing in the sewage treatment works.

Empty Fort Strategy: sewage treatment machines are operated without chemicals added, or some businesses extract fresh water from underground directly to fake the scene that the water is treated so that they can hide from the inspections of the authorities.

Tunnel Warfare Strategy: some businesses build multiple drain outlets and connect them with blind tunnel as well as equip them with integrated control system so that the authorities can never find all of the drain outlets.

Mobile Warfare Strategy: some businesses carry the sewage water outside with oil tank trucks so that they can pour them to ditches or farmlands when nobody is at presence.

Strategy of Cross the Sea under Camouflage: some businesses boast through the media that they have achieved "zero emission" when they are discharging the sewage sneakingly. Some of them even bragged themselves as environmental protection enterprises, the technology of environmental protection of theirs ranking first in this country even in the world, by which they divert the public's attention.

Diversion Strategy: some businesses trickily discharge sewage water at night rather than in the day time, in rainy days rather than in shiny days and after the inspections rather than before the inspections.

...

People who suffered deeply from the pollution didn't think that they still have much curiousness and hope because they, the farmers, fishermen, boatmen, and citizens, had already seen too many journalists for ten years. They felt that reports cannot help them.

In spite of this, Huo Daishan didn't give up. As a person who is raised up by

At some places, sanitary sewage is directly discharged into river through underground pipelines.

Ecological Civilization of Contemporary China

Huaihe River, he had a very strong sense of responsibility and he adhered to use his camera to record the real situation concerning the pollution in Huaihe River and the living environment of people lived along it.

Links

- In the middle of July 1994, after a rainstorm hit the upper stretches of Huaihe River, the level of Yingshang Reservoir rose to exceed the warning line. The Reservoir was therefore opened to discharge, then 200,000,000m3 sewage accumulated for the whole winter and spring flowed to the middle and lower reaches of Huaihe River. And it finally caused the "Huaihe River water pollution event" to shock the whole world.

- On August 8, 1995, the State Council issued the first regional water pollution control law *Interim Regulations Concerning the Prevention*

Workers drive the water clean keeping boat, clearing the household garbage on the river.

and Control of Water Pollution Within the Territory of the Huaihe River Valley. On June 29, 1996, the State Council approved to issue Plan for Prevention and Control of Water Pollution Within the Territory of the Huaihe River Valley & The Ninth Five-Year Plan, and laid emphasis on prevention and control of water pollution within the territory of the Huaihe River Valley by including it into "three Rivers and three Lakes" program of the ninth Five-Year plan.

- On January 11, 2003, since the goal of 1996 Plan hadn't been achieved, the State Council gave an official reply on The Fifteenth Five-Year Plan for Prevention and Control of Water Pollution Within the Territory of the Huaihe River Valley to establish the goal: total amount of discharge of the main pollutant COD shall be deducted from 1.059 million tons per year in 2000 to 643,000 ton per year in 2005, while total amount of discharge of ammonia nitrogen shall

"Huaihe River Safeguard" Huo Daishan shows people in the street of Haozhou City, Anhui the Huaihe River protection pictures he shot.

deducted from 152,000 ton per year to 113,000 ton per year.

- From May 28 to June 11, 2004, SEPA sent three inspection groups to investigate the situation of the key industrial pollution sources, urban sewage treatment works and the key cross sections of Huaihe River in 91 districts and counties, 21 cities of four provinces along the Huaihe River. 165 key industrial pollution sources related to chemical, brewing, pharmaceutical, paper and other industries, 30 municipal sewage plants, 65 rivers (lakes and sewage ditches) and 155 cross sections of river were included in this investigation.

The most impressive thing in the pollution control of Huaihe River is not how much money the government has invested but the efforts given by all volunteers. And the most memorable thing is not the shocking pollution but the astonishing "Thirty-Six Strategies" spread around.

Beautiful China

For a beautiful China, its beauty must be human-related and environment-oriented.

Real beauty and poetic inhabit place can only be found where life is in bloom.

Without trees, especially trees in their natural status, home of mankind would be far from livable, let alone poetic no matter how modernized and luxurious the home is. Is concrete road poetic? Is "concrete forest" poetic? How can it be poetic if bird's nest is nowhere to be found? When natural things are changed, spring gets more and more silent and such familiar vegetations as green bristlegrass are becoming traceless, even though the streets are bordered by outlandish flowers, our home would still be uninteresting however we decorate it.

Ecological Civilization of Contemporary China

Real modern life of high quality lies in the integration of soul and Nature and the harmony between life's pace and the rhythm of Nature rather than the number of property and money. In the view of daily life, such a state is a sort of demonstration of "unity of man and Nature". Once we realize it, the green bristlegrass on the lawn outside the window will bring us better sense of happiness, beauty of life and wonder of the world than potted valuable plants in the windowsill do.

Also because of this, environmental protection and building up beautiful China are not just confined to paying close attention and bringing changes to polluted lands and rivers, hunted natural creatures, poisoned air, grassland desertification, deforestation, cities filled with concrete buildings, commercialized resources and so forth. To ordinary people, they are better reflected in how life has been changed in a vicious way and how it will achieve virtuous cycle bit by bit. For example, in a city like Beijing, we

Green, a color lets us feel an exciting dream.

On April 26, 2014, the "Low Carbon • Healthy Family Life" family environmental protection promotion activity was launched in Chaoyang Park in Beijing.

should be able to touch naturally-grown trees, hear intoxicating warble of wild birds and appreciate the symphony produced by insects in the autumn.

Then, you will understand every story we tell, be it story of people or story of Nature, is actually your own story or is going to be your own story—you can do whatever the people in the story have done, for enjoying and creating beautiful environment is every people's natural right. The most important and fundamental thing for us to perform the right is to care for the stories of Nature, let them be part of us. After all, their stories are in fact our stories.

Just like us, they are creation of Nature, and they are named blue sky, green land, clear water…

Green Development:
Green Strategy of 3D Printing

The working process of 3D printing is like a green magic show:

A delicate picture faintly shows up after a bundle of red rays rapidly scanning on a layer of evenly-spread white powder, and the picture shows up again with another bundle of red rays scanning on anther layer of evenly-spread white powder.

The white powder used in this magic show is actually a kind of resin material that will be solidified by laser scanning under control of computer, and the delicate picture will finally be presented when seamless connection among layers is accomplished by the virtue of laser after times of scanning and solidification.

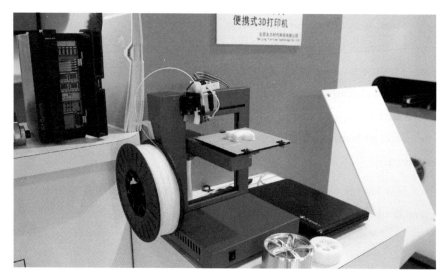

The first portable 3D printer worldwide.

Reciprocated overlays will finally deliver the three-dimensional components well designed beforehand – this is 3D printing.

The whole process of production is a little bit like building blocks, placing materials layer by layer until works finally delivered, and this way of production barely waste any material, cost-saving and environment-beneficial.

From the aspect of environmental protection, the prospect of 3D printing is very exciting.

3D printing needs neither paper nor ink. Instead, it makes specific metal powder or memory materials melt and overlaid layer by layer based on instructions of electronic model figure and eventually turns the electronic figure into real object through series of techniques like electronic drawing, remote data transmission, laser scanning and material melting.

3D printing is a technology revolution in global industry. It will significantly lower production costs and reduce environmental pollution, which is also the long-cherished wish for manufacturing industry in China to achieve green upgrades.

3D printing technology has wide range of applications in China, mainly in three aspects at present. Firstly, it can apply to high-end products, such as aerospace industry that require relatively high precision, possess complex structure and yields in very small amount of output. Secondly, it can apply to mid-end industrial products, such as in developing new products, when samples are usually needed at each stage of development and design and 3D printing technology provides a relatively quick and economical solution. Thirdly, it can also apply to products for civil use, such as in the medical field. For example, when printing artificial joints, 3D technology can realize customized printing according to joint size of specific person.

The dedicated space cushion, fax shell, casting wax of automotive gasoline engine, precision castings with complex modeling of Yang Liwei, the well-known astronautic hero, are all modeled with 3D technology and ultimately manufactured into finished products.

In China, the overall volume of 3D market printing is relatively small with decentralized enterprises and limited social influence due to lack of support on original core technology and material resources. In 2011, 3D equipment ownership in China accounted for only 8.6% of global inventory, with huge distance from Europe and other countries.

In January 2013, the research on rapid prototyping by Professor Wang Huaming from Beijing University of Aeronautics and Astronautics won the First Prize of State Science and Technology Invention Award, giving confidence to the Chinese market.

Now you can imagine: for example, it is no longer a dream to get one pair of shoes that are completely fit and consistent with your own personalized design. Such shoes do not need you to shuttle between numerous malls to find out like a Chinese saying of "looking for something by wearing out iron shoes", and with the help of 3D printing technologies, you can print out shoes that not only completely fit your feet, but also can be designed with preferred surface pattern and tread groove on the computer.

Of course, what' more important is that the production of such shoes leaves no waste and source of pollution.

Links

- 3D printing, also known as additive manufacturing, refers to the way of printing rather than processing all types of daily and industrial products. It is a method of direct manufacturing real components by accumulating materials (usually material pileup layer by layer) based on three-dimensional electronic model, which can achieve 100% of utility rate of material without mold.

The first batch of 3D printing houses debuting in Shanghai.

- In January 2013, on CES Exhibition held in Las Vegas, dabbed as "Electronics Feast of Global Consumers", 3D Systems exhibited 3D printers specialized for household market for the first time, providing the opportunity for 3D printing into the household market.

- In April 2013, Frost & Sullivan published 3D Printing Global Market Research Report in 2012, and the data showed that the global market size of 3D printing in 2010 was USD 1.325 billion, followed by a substantial growth in 2011 to USD 1.714 billion, an increase up to 29.36%.

- In December 2012, Su Bo, Vice Minister of the Ministry of Engineering and Information Technology, said in public that it would promote the industrialization of "3D Printing" and develop a roadmap and medium and long-term development strategies with enhanced finance and tax policy guidance.

- In January 2013, China's first 3D printing experience pavilion opened in Industrial Design and Creative Industry Base in Beijing. In the demonstration, a three-dimensional portrait was "cloned" and a customized three-dimensional blueprint was printed as a true object, which symbolized that cool 3D printing technology has become a reality.

Ecological Civilization of Contemporary China

The first 3D "picture making" store in Nanjing, providing high-precision human picture printing service.

- On April 26, 2013, the Ministry of Science and Technology issued *Guide for National High-technology Research and Development Program (863 Program), National Science and Technology Supported Program in Manufacture Industry – Alternative Projects Collected for 2014*, in which 3D printing industry was first listed. The Guide mentioned that the project is funded with a total amount no more than RMB 40 million for research.

3D printing has evolved from concept to the thing that gradually changes our lives. China, as one of largest manufacturing countries in the world, urges to make green strategy of 3D printing widely applied to various industries in order to reduce the negative impact on environment, on the other hand, it has to balance the problems of renewable energy supply and raw material inputs against the application of this new technology.

Cyclic Development: The Lights-out Earth Hour

At 6:30 p.m., March 28, 2009, the Bird's Nest and the Water Cube were crowed with people.

The Bird's Nest was the first to turn off the lights. "10, 9, 8...3, 2, 1", the red and yellow lights of the top started to go out amid the countdown, closely followed were the lights in the middle and bottom area, the whole process took less than 10 seconds.

In contrast to the staged lights out in the Bird's Nest, the blue lights of the Water Cube went out completely in an instant. Cheers and a lasting round of applause broke out in the darkness.

This is part of the event called "Lights-out Earth Hour" held in Beijing on March 28, 2009.

Meanwhile, the lights-out event circled the earth by starting from Chatham Islands on the eastern coast of New Zealand then entering into Asia through Oceania while starting from Sydney Opera House in Australia to Las Vegas casinos in the United States and from Great Pyramid of Giza to the Eiffel Tower in Paris...and finally ended in North America. At the moment, over 3000 cities from 84 countries and regions across the globe turned off all the lights for an hour, appealing to energy conservation and green house gas emission reduction with concrete actions.

As the country representing the fifth time zone in the lights-out event, China proceeded with this global lights-out relay by engaging its various major cities in this event including Beijing, Shanghai ,Hong Kong, Macau, Baoding, Dalian,

Ecological Civilization of Contemporary China

Nanjing, Shunde, Hangzhou, Changsha, Harbin and Changchun, showing its resolution in addressing the problem of global warming with joint efforts.

The "lights-out" action carried out in 2009 was the first organized large-scale engagement by China in the "Earth hour" event. According to the real-time power load monitoring system of Beijing, the power load of the moment within Beijing when the Beijing action started was reduced by around 70,000 kW compared with its normal level.

Compared with the maximum power load of 12.48 GW across Beijing in 2008, 70,000 kW is only a small change in number but which still well reflects the public's concern over energy conservation.

The Beijing action of the "lights-out" event went along. At 8:29 p.m., the logo of a land-marking mall in Beijing and the neon lights as well as the landscape lights around it went out instantly, turning the normally blazingly

Lights in the Bird's Nest in Beijing are off to respond to the "Earth Hour" activity.

colorful building into a plain object. During the time when the lights were out, groups of volunteers held candles to form a pattern of "60" as their response to the Beijing action of the "Lights-out Earth Hour".

As residents on earth, the Chinese public's sense of environmental protection took a "skyrocket" outburst. This kind of powerful and distinct outburst represents the sublimation and awakening of spontaneous rise of their sense of environmental protection.

Although it is impossible to conduct statistics on the amount of energy the "Lights-out Earth Hour" event has saved yet, the demonstrative effect of the concept of environmental protection it generated has proven to be far greater than regular preaching.

As Li Bingbing, the Chinese ambassador of the "Lights-out Earth Hour" once put it, the event sends a strong signal to everyone that a minor action can actually change the whole world and everybody can contribute to the undertaking of global warming alleviation.

A volunteer on site expressed his support for this kind of activities who not only turned off the lights of his own house but also called for over 10 friends to join him in this action.

Links

- World Wildlife Fund (WWF) is one of the world's largest independent environmental NGOs. Since its founding in 1961, WWF has been concentrating in the environmental protection effort, with around 5.2 million supporters around the globe and websites in over 100 countries. WWF endeavors to: protect the biological diversity of the world, ensure the sustainable utilization of the renewable natural resources and promote activities focusing on the reduction of pollution and wasteful consumption.

Ecological Civilization of Contemporary China

The globally set advertising lamp boxes in walkway of Shanghai Subway.

- "Earth Hour" is a global initiative brought up by WWF in 2007, which calls for individuals, communities, enterprises and governments to turn lights off for an hour on the last Saturday in March every year, in an effort to show their support for the joint efforts in resisting global climate warming by inspiring the sense of responsibility for earth protection in the public and encouraging them to reflect on environmental issues such as climate change.

The first round of "Earth Hour" event was carried out in Sydney, Australia on March 31, 2007 and participated by over 2.2 million families and enterprises. Soon afterwards, the event swept the globe at an amazing speed.

In 2012, 124 Chinese cities joined the "Lights-out Earth Hour" event held by WWF and made public their environmental protection commitment.

Beautiful China

"Earth Hour" Chinese promotion ambassador Li Bingbing.

Meanwhile, over 1400 enterprises have enrolled in the "Lights-out Earth Hour" event nationwide.

WWF initiated the "Green Week" campaign in China in 2013 where the general public were invited to celebrate this environmental festival with concrete actions: choosing one day out of the seven days—each day representing a different green activity in a week and sticking to the following plan during 2013:

Monday to eat more vegetables; Tuesday to use environment-friendly bags; Wednesday no driving; Thursday to bring your own chopsticks, Friday no leftovers; Saturday to express your love for animals; and Sunday to go outside.

Ecological Civilization of Contemporary China

College students in Chengdu light the candles to pray for the earth in the "Earth Hour" activity.

What can we do to slow down global worming? Global warming seems to be a global concern, but it is also closely related to every one of us. In fact, environmental protection is as simple as starting from me for just one hour.

Low-carbon Development: Green GDP of Hezhang

The battle between the sun and the dark clouds was always on in the sky of Hezhang.

Since 1980s, the "indigenous zinc smelting" had been strongly advocated under the slogan of developing township enterprises. Spelter output of Hezhang County had ranked among top five nationwide, accounting for one sixth of the country's total output.

By 1990s, the market price of refined zinc here had reached tens of thousands of yuan per ton.

Under such a background, the entire county started to light the blazing fire for "indigenous zinc smelting" at any risk.

Every nightfall, flames on the furnaces could be seen everywhere in villages and towns along the National Highway 326, including Pingshan, Yemachuan and Magu. It was very splendid that flames lighted the villages and smoke enveloped the households.

With billowing smoke and flames, it was like that the "creepy monster" was about to come out of the cave.

Taking Magu Town, the largest town in Hezhang County, as an example, although the population was less than 40,000, the town produced an annual output of spelter of 18,000 tons with the value reaching more than RMB 400 million. These data could even shock the "creepy monster".

Hezhang had really gone crazy!

Why crazy? Staff of the environmental protection agency did some accounts once: a hexagon furnace which was fed with zinc ores of a grade of 40% could produce 0.8 ton of spelter per day which could bring an income of RMB 25,000 during the period from 2004 to 2005; with 20 furnaces of output per month, the profit could reach RMB 0.3 million; deducting the investment cost, one could be a millionaire in one year.

Located at the foot of Jiucaiping, the highest mountain in Guizhou, Hezhang County was once a designated poverty-stricken county of the country. There was a folk song going that "Buckwheat is what people live on at Jiucaiping and only during the confinement after childbirth can one have a corn meal". The County was haunted by a bad reputation that "it is a place that one doesn't want to go", which means that natural conditions there are hostile and poverty is severe and covers a wide scope.

The difference is quite huge. In 1990s, one could become a millionaire in only one year. Villagers joke that even the "creepy monster" can change his profession, who won't go crazy?

Jiucaiping Mountain in Xingfa Township, Hezhang County, Guizhou Province.

With the madness of Hezhang, the sky, earth and water started to go wrong-

According to a general survey of lead poisoning phenomena in 1994, the population suffering from 1-3 grades of lead poisoning accounted for more than 30% and lead content in blood of the sufferer was more than a dozen times and even a hundred times of that of healthy people. In 1997, as small mines were exploited and deforestation was carried out throughout the year, the mountain became loose which together with significant block of the river ways caused enormous mountain torrents that killed 12 people. As toxic substances such as waste residue of zinc smelting entered into the river ways with the help of rainfall, cancers and pneumoconiosis took place frequently...

Who is more vicious and crazier? The bleeding wound made Hezhang to calm down.

There is no way back and only one way presents itself in front of them that replacing the "black" with the "green" in no hesitation.

Hezhang County has launched the "three one-million programs" to build a specialty industry demonstration area, i.e. it is expected that by 2015 the sheep inventory of grassland ecological animal husbandry of the whole county will reach one million, the planting area of walnuts will reach one million *mu* and that of traditional Chinese medicine will also reach one million *mu*.

At the local tourism industry development conference held in October 2011, the unit price of "Hezhang walnut" produced by a walnut tree of more than 160-year old and about 16m high and is called the "King of Walnut" reached a high price of RMB 2400.

Now, the "creepy monster" has chosen a low-carbon green way and he has done such accounts:

A potful of smashed maize flour and fermented grist that is filtered, re-boiled and added with some walnuts can earn 200-300 yuan. Who does it knows

Ecological Civilization of Contemporary China

well. "Let's sell walnut candy!"

Links

- There are 33 kinds of metal and non-metal ores in Hezhang County among which 13 varieties have large reserves and high grade. The prospective reserves of the coal is about 4 billion tons while that of the iron ore is 405 million tons, accounting for 40% of the proved reserves of the whole province. The reserve of lead zinc ore ranks the first in Guizhou with that of germanium ranking the first in Asia.

- In 2006, Hezhang County established a leading team of banning the "two local methods". Led by heads of four major governmental offices and participated by heads of more than 20 functional departments including the environmental protection department, public security department, land and resources department and power supply

A peasant woman herding cattle in Hezhang County, Guizhou Province.

department of the County, 589 furnaces for zinc smelting by local method, 78 indigenous coke ovens and 65 zinc pot factories were clamped down. Meanwhile, more than 2000 illegal mines were exploded and closed with 58 small iron smelting furnaces, 4 improved coke factories and 1487 furnaces for zinc smelting by local method being eliminated and 32 coal wells being consolidated.

- Since 2009, Hezhang County has invested approximate RMB 200 million in the development of walnut industry. The planting area of walnuts has reached 460,000 *mu* with that of the walnut trees bearing fruits totaling 100,000 *mu*. The annual output of nuts has exceeded 7500 tons with the value reaching RMB 300 million. For walnuts in Hezhang, such honors as the "Hometown of Walnuts in China", "Recommended Fruit for Olympics" and "Top Ten Famous and High-quality Walnuts in China" have been obtained.

The highway to Jiucaiping Mountain, Jiulong Village, Hezhang County, the ridge of Guizhou.

Ecological Civilization of Contemporary China

Beautiful Yi girls in Hezhang, Guizhou Province.

- In 2012, the total forest area of Hezhang has reached above 2 million *mu*, increasing by approximate 800,000 *mu* compared to 1988. The forest coverage rate has reached 37.26% and the greening rate has reached 47.29%, increasing by about 15 and 17 percentage points respectively compared to 1988.

- From 2006 to 2010, total output value of Hezhang increased from RMB 1.672 billion to RMB 3.7 billion, an average annual growth of 13.46%. The general financial revenue rose from RMB 152 million to RMB 540 million, an average annual rise of about 23%. The urban disposable income increased from RMB 5982 to RMB 14,157, an average annual rise of 18.71%. The rural per capita net income grew from RMB 1747 to RMB 3126, an average annual rise of 12.34%. The total output of grain of the County achieved 944,400 tons with the planting area reaching 5.8464 million *mu* and the total value of agricultural output reaching RMB 3.69343 billion.

To have clean water and blue sky and transfer from "black" to "green", we need to start from the adjustment of industrial structure of green agriculture which helps actively explore new ways, methods and mechanisms to popularize and implement low-carbon green projects.

The low-carbon green GDP, a development state created and shown to meet the requirements of environmental protection and health, is now demonstrating an upward trend in the process of building a beautiful China.

Postscript

All Dreams Should Have a Beginning

—Story of Mr. Liang Congjie

Mr. Liang Congjie, whose grandfather was Liang Qichao and parents were Liang Sicheng and Lin Huiyin, was born in 1932 and died in 2010. He served as the member of the National Committee of CPPCC, the member of the Standing Committee of CPPCC, the member of the Subcommittee of Population, Resources and Environment of CPPCC, and the Founder and Chairman of the environmental NGO—Friends of Nature (FON). In 1999, Mr. Liang Congjie won the Earth Award jointly granted by China Forum of Environmental Journalists and Friends of the Earth Hong Kong and the Panda Award from the State Forestry Administration. He died at 4 p.m. on October 28, 2010 in Beijing, aged 79.

Ghostwritten Postscript

Liang Congjie, the first sponsor for folk environmental protection organization in China, and the founder and Chairman of "Friend of Nature" association.

I met with Mr. Liang Congjie for the first time on November 19, 2006 at the memorial of our friend and an animal protection volunteer—Wang Pei. In this memorial, I delivered a speech entitled In Memory of Wang Pei - You Run on the Ground Silently like a Swan Flying in the Cloud. After many years, I still can remember it clearly that, Mr. Liang Congjie, sitting next to me, swept for several times when I was giving the speech. The scene of him taking off his glasses and wiping tears is still vivid in my mind.

After exchanging business card with him, I was surprised that though his card was made from waste paper, a few lines of words printed on the card—Chairman of FON, member of the National Committee of CPPCC and History Professor—were quite impressive and aroused my admiration for this compassionate old man. Interestingly, a journalist who once interviewed him, was also impressed by his card as I did:

"Firstly, his business card drew my attention, which was the thinnest and the most rough card that I ever had seen. Bluntly, it was a piece of waste paper which can be seen everywhere."

"It is the most rough business card that I have ever seen and there is a production ingredients list on its back. He said that workers in the printing house cooperated with them well and they printed their business cards with waste paper.

Ecological Civilization of Contemporary China

It is hard to believe that Mr. Liang Congjie, as son of celebrities, actually led such a lifestyle."

Some years later, I successively got to know President of FON—Yang Dongping and Secretary General—Li Bo, whose business cards are also made of waste paper. In fact, this detail reflects FON's environmental protection principle of "wholeheartedly practicing what one advocates and never saying high sounding green words" and their practical actions.

For many years, Mr. Liang Congjie had always fulfilled this principle in all aspects of life. Apart from using business card made of waste paper, he also stored rinsing water produced by the washing machine to flush the toilet. In order to boycott disposable chopsticks used in the restaurant, he had always carried a small bag with several pairs of chopsticks in it for use of himself as well as his family members and friends. Even when attending the NPC and CPPCC, Mr. Liang Congjie would ride a bike. On one occasion, he went to attend the NPC and CPPCC by bike and consequently was stopped and inquired by the

Yang Dongping, Chairman and Director General of "Friend of Nature".

Ghostwritten Postscript

gatekeeper. He didn't stop riding until 2006 when his doctor suggested stopping doing so.

Mr. Liang Congjie considered that, though there is nothing to be said against a desire to have a better life, it is a crime against nature at the cost of destroying and infringing eco-environment. He said:" If you went to a restaurant with a pair of chopsticks several years ago, others would regard you as an idiot. Yet now, the waiters just smiles, which could be considered as an improvement. He said that "we shall neither say high sounding green words nor consider ourselves as the Green Savior. Instead we shall do it bit by bit."

Not only Mr. Liang Congjie but also a great number of members of FON have led an environmental friendly lifestyle. Everyone contacting with FON will learn that avoidance of disposable paper cups and half-cup-of-water-only are hospitality rules followed by staffs of FON. In addition, its internal publications are made of recycled paper. Such unity of knowledge and practice is inspiring more and more people. I find that many environmental NGO volunteers get used to carrying their own chopsticks and cups and they attempt to collect meeting name badges for recycle in any meeting they attend.

I have always believed that, in addition to a wide participation base of environmental protection, focus on every aspect of our life is extremely important. Therefore, what environmental NGOs in China shall endeavor to create and promote is exactly this new life movement with a characteristic of simplest action and focus on details of environmental protection. Once all people in China participate in such movement, it will change China's history and the world's history. The advocacy of this basic environmental protection concept and the exhortation for environmental protection actions shall become the future endeavor direction of NGO's environmental protection efforts in China.

Therefore, despite those tremendous deeds achieved by FON in environmental protection since its establishment in March 1994, such as saving rhinopithecus bieti, removal of Shougang Group, defending Nujiang River and protecting Tibetan antelope, and honors won like Asia Environmental Award, Earth Award, Ramon Magsaysay Award, what I value most is its detailed environmental protection idea and action enlightenment education with a characteristic of "simplest action". As the enlightenment to Chinese people for their environmental protection awareness, FON does well in this regard but there is still a huge room for improvement.

On this point, Mr. Liang Congjie had a clear understanding. He once said that though no one could be the savior, there is hope only when everyone behaves herself or himself. Jane Goodall, an internationally famous environmental protection activist and expert on behavioral study of chimpanzees, is a friend of Mr. Liang Congjie and she also values individual's role in environmental protection. She said that though a single person can do little, we are able to change the world if we join our hands.

As a result, though it is established 3 years later than the environmental NGO- Liaoning Panjin Saunders' Gull Conservation Society, FON becomes the

NGO volunteers participate in environmental protection activities.

most typical environmental NGO in China beyond any controversy. So far, it has developed and accumulated its membership up to more than 8000 people and cultivated a group of leaders of new environmental NGOs, such as the convener of Green Earth Volunteers- Wang Yongchen.

Ji Xianlin, a scholar mogul, once commented on Mr. Liang Congjie's devotion to environmental protection as follows:" As a historian, if Mr. Liang Congjie goes on in this way, he will have success without risks. However, he is unwilling to stay in the ivory tower and live comfortably; instead, he firmly gives up a road to success without any disaster or difficulty and transforms himself from a historian to a friend of nature, which reflects his care for the fate of his nation and his people and meets the wishes of the people and follows the trend. I can only show my admiration and respect for him. I would rather have a friend of nature than have a historian."

Why did Mr. Liang Congjie give up history study to devote to environmental protection and found the environmental NGO- FON? How did his dream for environmental protection begin? What is the inspiration for the public? There is actually a story accountable for these questions.

In terms of this story, Southern Weekend once recorded as follows:

In the spring of 1997, Yang Dongping, the representative of FON, visited the New York-based Environmental Defense Fund (EDF), which was a celebrated environmental NGO established in 1967 with 300,000 members. The founder of EDF told that there were only 10 people at first who gathered in the sitting room and considered that they must do something. "I can't help smiling. All the dreams begin at the same point. Originally we had 4 people and we sat in the park to consider."

The dream began on June 5, 1993, a day when around 40 intellectuals gathered on the grass lawn of Linglong Park to discuss severe environmental conditions in China with grave concerns under the shadow of desolate ancient tower near Beijing. It was a spontaneous assembly without name, meeting place, agenda and media involvement. Mr. Mr. Liang Congjie said that they later named the assembly "Linglong Park Assembly" after name of the park.

Before then, four of them prepared to set up an environmental NGO in China. They were Mr. Mr. Liang Congjie, as the member of the National Committee of CPPCC and tutor of International Academy of Chinese Culture; Yang Dongping, as the researcher of Institute of Higher Education of Beijing Institute of Technology; Wang Lixiong, an independent writer and explorer; and Liang Xiaoyan, as the editor of Eastern Miscellany.

In the Linglong Park Assembly, a member's speech greatly moved everybody: if 1.2 billion Chinese aim to reach a living standard of Americans, resources needed will be 60 times that of today according to planning and management experts. How can I believe that the land breeding us for 5000 years is able to provide 60 times of resources?

"Green Earth" little volunteers spreading environmental protection knowledge in the street.

"With the greatest population devouring the least resources, China's future is facing a double danger." This sentence deeply stings these intellectuals.

At that moment, Mr. Liang Congjie and his friends realized that they should begin to take concrete measures to protect the environment. The idea of establishing an environmental NGO came across their minds. Several years later, Mr. Liang admitted that he was not sure about what to do exactly at that time, yet, he wanted to make some contributions to China's environmental protection.

Yes, all dreams should have a beginning. It occurs to me that every Chinese can have such a dream. Every one may, starting with their own dream, take actual actions to make his/her lifestyle and life greener, the environment greener, China greener, and the world greener.

In the opinion of Yang Dongping, the second Chairman and incumbent President of Friends of Nature, one of the most gratifying and inspiring successes of the organization is its great demonstrative role and diffusion effect. Nowadays, members of Friends of Nature across China are actively carrying out a variety of environmental protection activities in cities from Beijing, Shanghai, Shenzhen, Chongqing, Nanjing, Wuhan, Xi'an, Shijiazhuang, Xiangfan to Hoh Xil and Shangri La, etc. Although most of their petitions to protect nature unheard, they maintain a positive attitude of "challenging the impossible" and are stepping up their efforts in spite of previous failures. They are a group of selfless challengers, sober devotees and optimistic "pessimists". Thus, people often deem them naïve or extreme. Nevertheless, the society does not always follow the will of the "smart people". It's such a blessing when they look back to find that we've achieved more than we'd expected.

Since the implementation of the reform and opening-up policy, NGOs in China have been growing steadily. Whereas, they are still small in number and slow in growth on the whole. Up to the end of 2010, China had 446,000 NGOs, an increase of 15,000 from 431,000 in 2009. The growth rate, yet, dropped 0.6% to 3.5% in 2010, compared with 4.1% in 2009. As with the development of

"Green Earth" Reporter Salon in 2005.

political civilization, adequate development of NGOs and the third sector will lay a sound foundation for the sustainable development of society. If the first decade is the stage when NGOs in China start from scratch to sow its seeds, then the next decade will see NGOs grow and thrive. NGOs will shoulder more diverse missions and tasks and play a more important role in rendering social services and meeting social needs.

In this process, NGOs should not only break new grounds and put forward more challenges, but also diagnose and construct; meanwhile put forward questions as well as solutions. In view of sustainable development, the scope of environmental protection should be naturally expanded to community and rural areas, as without improvement of basic living environment of the public, ecological and environmental protection in its real sense is out of question. Certainly, the prerequisite is the self-building and sustainable development of

NGOs. The tremendous developing opportunities facing NGOs also mean severe challenges ahead.

Mr. Liang Congjie's efforts in environmental protection set a very good example. The future of China's environmental NGO dream is surely promising as well.

Accordingly, with China's environmental NGO dream, Chinese's dreams on an eco-China and beautiful China are sure to be brilliant as well. Though Mr. Liang Congjie has finished his lifetime journey of pursuing his dream on the vast land of China, one after another Chinese, following his steps, are making endeavors to start their journeys and live their dreams on the same land with their new life.

When we enjoy the blue sky, green land, and clear water many years later, I believe we will cherish the efforts we are making at present. By that time, I believe Mr. Liang Congjie's face will broaden with his warm smile in Heaven

NGO volunteers were telling stories to pupils in Qiunatong Primary School (Nujiang, Yunnan) in 2006.

Ecological Civilization of Contemporary China

NGO volunteers gathering in the highest peak of Great Khingan.

when he sees the "ecological red line" has been converted to the "happy green line". Therefore, we should spare no efforts to make the sky bluer, land greener and water clearer.